Philip Atkinson

The Elements of Dynamic Electricity and Magnetism

Second Edition

Philip Atkinson

The Elements of Dynamic Electricity and Magnetism
Second Edition

ISBN/EAN: 9783744790420

Printed in Europe, USA, Canada, Australia, Japan

Cover: Foto ©berggeist007 / pixelio.de

More available books at **www.hansebooks.com**

THE ELEMENTS

OF

DYNAMIC ELECTRICITY AND MAGNETISM.

BY

PHILIP ATKINSON, A.M., Ph.D.,
AUTHOR OF "ELEMENTS OF STATIC ELECTRICITY" AND "THE ELEMENTS OF ELECTRIC LIGHTING."

SECOND EDITION.

NEW YORK:
D. VAN NOSTRAND COMPANY,
23 MURRAY AND 27 WARREN STS.
1892.

COPYRIGHT, 1891,
BY
D. VAN NOSTRAND COMPANY.

ROBERT DRUMMOND,
Electrotyper and Printer,
New York.

INTRODUCTION.

THIS book was written for learners rather than for the learned. Previous to the last decade the demand for electric books was confined chiefly to scientific investigators versed in the higher mathematics, and the authors of such books were electricians of the same class, who recognized the importance of mathematical accuracy in treating electric phenomena. Hence mathematical formulæ became a prominent feature of such books. But the various electric industries to which the recent unprecedented electric development has given rise, have given employment to a numerous class of persons to whom mathematical books are almost unintelligible, and yet to whom a scientific knowledge of the various kinds of electric apparatus which they are required to operate, or with which their business is connected, is of the highest importance. There is also a class of liberally educated persons who desire to extend their knowledge of electric principles, but have not the time or patience to follow the intricacies of mathematical formulæ, especially in the abbreviated form usual in the books referred to. A third class are students who intend to become electrical engineers, to whom a thorough knowledge of elementary, physical, electric principles is important as a preparation for a

Company, the Electrical Supply Company, Central Electric Company, the Electrical Accumulator Company, "C & C" Electric Motor Company, the Western Union Telegraph Company, the Chicago Bell Telephone Company, the American Telephone and Telegraph Company.

PHILIP ATKINSON.

CHICAGO, November 1, 1890.

CONTENTS.

CHAPTER I.

THE VOLTAIC BATTERY—DEFINITIONS. 1
　Dynamic Electricity Defined. Discoveries of Galvani. Discoveries of Volta. The Couronne de Tasses. The Voltaic Pile. Value of Volta's Discoveries. Cell, Element, and Battery. Battery Sign. Electrodes and Poles. Conditions of Electric Energy. Electromotive Force. Resistance. Current. Units of Electromotive Force, Resistance, and Current. Operation of the Voltaic Cell. Theory of Electric Generation in the Cell. Amalgamation of the Zinc. Insulation and Clamping. Polarization.

CHAPTER II.

ONE-FLUID CELLS. 13
　Smee's Cell. Zinc-Carbon Cells. Walker's Cell. Potassium Bichromate Cell. The Grenet Cell. The Mercuric Bisulphate Cell. The Leclanché Cell. The Law Cell. The Diamond-Carbon Cell. Dry Cells. Polarization of One-Fluid Cells.

CHAPTER III.

TWO-FLUID CELLS. BATTERY FORMATION. 23
　Construction of Two-Fluid Cells. The Daniell Cell. The Callaud Cell. The Grove Cell. The Bunsen Cell. The Silver Chloride Cell. Battery Formation. Connection between Cells.

CHAPTER IV.

MAGNETISM. 35
　The Natural Magnet. Magnetic Polarity. The Mariner's Compass. The Surveyor's Compass. The Earth's Magnetic

Poles. Declination. Inclination or Dip. The Dipping Needle. Magnetic Maps. Terrestrial Magnetism Illustrated. Magnetic Intensity. Magnetic Force Ascertained by Oscillation. Magnetic Force Ascertained by Deflection. Absolute Magnetic Intensity. Biot's Law. Origin of Terrestrial Magnetism. Secular Variation. Secular Variation in the United States. Annual and Diurnal Variation. The Eleven Year Period. Magnetic Storms. Cosmic Variation. Exact Observation. Secular Variation at Washington. Secular Variation at San Francisco. Artificial Magnets. Magnetic Saturation. The Armature. Laminated Magnets. Magnetic Loss. Portative Force. Polar Attraction and Repulsion. Magnetic Lines of Force. Magnetic Field. Form of Magnets. Magnetic Penetration. Location of the Poles. Paramagnetic and Diamagnetic Bodies. Magneto-Crystallic Induction. Magnetism as a Mode of Molecular Motion. Analogy between Magnetic and Electric Phenomena. Coulomb's Torsion Balance. The Gauss-Weber Portable Magnetometer.

CHAPTER V.

ELECTROMAGNETISM. 71

Deflection by the Electric Current. The Galvanoscope. The Schweigger Multiplier. Ampère's Rule. The Astatic Needle. Compensating Magnet. Cause of Deflection. The Electromagnet. Electromagnetic Poles. Winding. Magnetic Strength. Core. Helix Coefficient of Magnetic Induction. Electromagnetic Saturation. Form of Electromagnets. Armature. Experiments in Diamagnetism. List of Diamagnetic and Paramagnetic Substances. Deflection of the Electric Current by the Magnet. Ampère's Table. The Solenoid. De La Rive's Floating Battery. Mutual Induction of Electric Currents. Rotary Movement by Current Induction. Ampère's Theory of Magnetism. Generation of Electric Currents by Induction. Current Induced by Magnet. Current Induced by Another Current. Current Induced by Opening or Closing Primary Circuit. Current Induced by Varying the Strength of Primary Circuit. Results of Current Induction. Generation of Current Dependent on Variation of Intercepted Magnetic Force. Coefficient of Mutual Induction. Self-Induction. Extra Current. The Spark. In-

CONTENTS. ix

duction of Core. Induction Coil. Condenser. Interrupter. Sliding Core. Water Rheostat. Construction of Core. Operation of Condenser. Leyden Jar as a Condenser. Special Construction. Ruhmkorff's Commutator. The Coil a Converter. Electric Perforation. Physiological Effects of Faradic Current. Discharge in Air and in Vacuo. Electric Gas Lighting. Spark Coil.

CHAPTER VI.

Electric Measurement. 110

Electric Potential. Electromotive Force. Electric Resistance. Insulation and Conductivity. Electric Current. Ohm's Law. Electric Units. The Volt. The Mircrovolt. The Ohm. The Megohm. The Ampere. The Milliampere. The Ampere-Hour. The Coulomb. The Farad. The Microfarad. The Watt. The Electric Horse-Power. Different kinds of Electric Measurement. Electrometers. Galvanometers. Measurement of Angles. Angular Measurement of Deflective Force. Calibration of Galvanometer. Sine Galvanometer. Tangent Galvanometer. Astatic Galvanometer. Thomson's Reflecting Galvanometer. Differential Galvanometer. Ballistic Galvanometer. Common Galvanometers. Voltmeters and Ammeters. The Weston Voltmeter. The Weston Ammeter. The Weston Milliammeter. The Wirt Voltmeter. Ayrton and Perry's Spring Voltmeters and Ammeters. Gravity Ammeters. The Cardew Voltmeter. The Edison Current-Meter. The Forbes Coulomb-Meter. Voltameters. The Water Voltameter. The Weber-Edelmann Electrodynamometer. Measurement of Electric Resistance. Resistance Coils. The Wheatstone Bridge.

CHAPTER VII.

The Dynamo and Motor. 165

The Magneto-Electric Generator. Commutation. The Alliance Machine. The Siemens Armature. Wilde's Machine. The Dynamo. Ladd's Machine. The Pacinotti-Gramme Armature. Improved Commutator. Direction of Current. Interior Wire of the Gramme Armature. The Cylinder Armature. Closed-Circuit and Open-Circuit Armatures. Location of the Armature's Magnetic Poles. Magnetic Lag. Position of the Brushes. The Field-Magnets. Series, Shunt, and Compound Winding. Constant Current Dynamo. Constant Potential Dynamo. The Edison Dynamo. Alternating Current Dynamos. The Gordon Dynamo. The Westinghouse Dynamo. Separate Excitation. Advantages of the Alternating Current Dynamo. The Converter. Development of the Electric Motor. The Dynamo as a Motor. Principles of the Motor. Loss of Energy. Eddy Currents. Series, Shunt, and Compound Wound Motors. Reversible Rotation. The Alternating Current Motor. The Westinghouse Tesla Motor. The Tesla Motor as a Converter. Reversal of Rotation. Distribution of Power. Elevated-Road Distribution. Thermo-Magnetic Motors.

CHAPTER VIII.

ELECTROLYSIS. 206

Nomenclature by Faraday. Theory of Grotthus. Electrolysis of Water. Conditions of Electrolysis. Secondary Reaction. Electrolysis of Mixed Compounds. Relations of Electrolysis to Heat. Lowest Required Electromotive Force. Faraday's Laws. Magnetic Effects. Chemical Equivalence. Electrochemical Equivalence. Effect of Current Reversal. Effect of Convection. Relative Conditions of Current and Electrolyte. Electroplating. Various Details. The Anodes. Plating Solutions. Auxiliary Operations. Required Electric Energy. Required Time of Immersion and Thickness of Deposit. Agitation of the Solution. Electrotyping. Electric Refining of Metals. Electric Reduction of Ores. The Hall Process for Aluminium.

CHAPTER IX.

ELECTRIC STORAGE. 233

The Leyden Jar and Condenser. Grove's Gas Battery. Planté's Secondary Cell. Chemical Reaction. The Faure Cell. Chemical Reaction. Defects of the Faure Cell. Improved Faure Cell. Electric Preparation of the Plates. Electric Energy of Improved Cell. Effects of Charge and Discharge on the Plates. E. M. F. of discharge. Conductivity and Buckling. Weight of Cells. Composition of Grids.

The Julien Cell. The Pumpelly Cell. Durability of Storage Cells. Storage Capacity. Relative Time of Charging and Discharging. The Hydrogen Alloy Theory.

CHAPTER X.

THE RELATIONS OF ELECTRICITY TO HEAT. 252

Heat Developed by Electric Transmission. Joule's Law. Joule's Equivalent. Heat Developed by Electrochemical Action. Electro-Thermal Capacity of Conductors. Electric Blasting. Electric Cautery. Electric Fuses. Thermo-Electric Generation. Thermo-Electric Diagrams. The Peltier Effect. Thermo-Electric Inversion. The Thomson Effect. The Thermopile. Electric Welding.

CHAPTER XI.

THE RELATIONS OF ELECTRICITY TO LIGHT. 279

The Relations of Electric Heat to Electric Light. Photo-Electric Generation. Photo-Electric Reduction of Resistance in Selenium. Polarization of Light. Magneto-Optic Polarization—Faraday's Discoveries. Verdet's Discoveries. Becquerel's Discoveries. Kündt and Röntgen's Discoveries. Kerr's Discoveries. Effects of Double Reflection. Summary. Maxwell's Theory. Molecular Theory. Strain in the Media. Electric Lighting. The Arc Light. The Arc. Electric Candles. The Arc Lamp. The Crater and Point. The Heat and Light. Establishment of the Current. The Carbons. Automatic Regulation. Hefner von Alteneck's Regulator. Series Distribution. Automatic Cut-Out. The Incandescent Lamp. The Filament. Filament and Lamp Attachment. Position and Current. Parallel Distribution. Multiple Series and Series Multiple. Three-Wire System.

CHAPTER XII.

THE ELECTRIC TELEGRAPH. 310

Early History. The American Morse Code. The International Morse Code. Simple Line Equipment. The Battery. The Key. The Register. The Sounder. The Relay. Cut-Out, Ground-Switch, and Lightning-Arrester. Line

Construction. Station Arrangement. Switch-Board. Repeaters. The Button Repeater. The Milliken Repeater. Repeater Connections. Duplex Telegraphy. The Stearns Duplex. The Polar Duplex. The Pole-Changer. The Polaarized Relay. Operation of the Polar Duplex. Quadruplex Telegraphy. Construction and Operation of the Quadruplex. Repeating by the Quadruplex. Substitution of the Dynamo for the Battery. The Wheatstone System of Automatic Rapid Transmission. Submarine Telegraphs. Locating Faults. The Dial Telegraph. Printing Telegraphs.

CHAPTER XIII.

THE TELEPHONE 359

Early History. Principles of the Telephone. The Bell Telephone. Improved Transmitters. The Edison Transmitter. The Blake Transmitter. Accessory Apparatus. The Signaling Apparatus. The Exchange. The Multiple Switch-Board. Hughes' Microphone. Theory of Telephonic Transmission. Multiplex Telephony. Long Distance Telephony. Van Rysselberghe's System. The American System. The Hunning Transmitter. Transmission on Long Distance Lines.

THE ELEMENTS OF DYNAMIC ELECTRICITY AND MAGNETISM.

CHAPTER I.

THE VOLTAIC BATTERY. DEFINITIONS.

Dynamic Electricity Defined.—The term *dynamic*, from δύναμις, power, is appropriately used to designate electricity when employed for useful work, embracing the electric phenomena pertaining to that state of electric motion termed *current*, by which apparatus or machinery is operated, as distinct from that class of phenomena termed *static*, which pertains chiefly to electricity when stationary and not employed in this way. Hence it may be accepted as properly including all the various electric phenomena to which the terms *galvanic*, *voltaic*, *current*, *chemical*, *magneto*, and *thermo* have been applied.

Discoveries of Galvani.—In 1780, Galvani, a professor of anatomy at Bologna, Italy, observed certain muscular contractions in the limbs of frogs recently killed, produced by electricity generated by a frictional machine. He subsequently noticed similar contractions when the frogs' limbs were hung on an iron balcony by copper hooks in contact with the lumbar nerves. Placing a pair of them on an iron plate, and touching the lumbar nerves with a copper wire the opposite end of which was in contact with the plate, he reproduced the muscular movements. From this he inferred that the nerves and muscles were oppositely electrified, and that the muscular action was due to the establishment of a connection between them.

Discoveries of Volta.—Volta, a professor of physics at Pavia, Italy, having observed that the movements were produced by using a muscle in connection with two metals, inferred that they were due to the electricity generated by the contact of the metals when the damp muscle was placed between them, and that if the same conditions were produced in some other way, electric generation would follow. On this hypothesis he constructed, in 1800, the apparatus known as the *couronne de tasses*, or crown of cups.

The Couronne de Tasses.—This apparatus consisted of a series of cups or glasses, arranged in a circle, each containing a zinc plate and a copper plate partly immersed in a solution of salt in water, the copper of each cup being joined by a copper conductor to the zinc of the next cup, the fluid intervening between the two metals. Connection being made by a conductor between the copper of the first cup and the zinc of the last, strong electric effects were obtained, and the discovery excited great interest in the scientific world, as friction was the only means previously known of generating electricity.

The Voltaic Pile.—Volta subsequently invented a portable apparatus, intended for medical, electric treatment in hospitals, known as the *voltaic pile*. This apparatus consisted of a series of copper and zinc disks, arranged in a pile, with disks of cloth, moistened with a solution of salt in water, between each pair; the lowest disk being copper, the next zinc, and the next cloth; the same order being continued throughout the pile, so that the topmost disk was zinc. Connection being made between the top and bottom disks, as between the terminal plates of the couronne de tasses, similar electric effects were obtained.

This apparatus was also constructed with copper and silver coins. Water acidulated with sulphuric acid was also used instead of the solution of salt in water, both for the pile and the couronne de tasses.

Value of Volta's Discoveries.—These discoveries laid the foundation of the science of dynamic electricity, and Volta's apparatus is the type of all the batteries since constructed. The value and importance of his work become apparent when we consider that after nearly a century of constant experiment by eminent scientists the metals he employed are still found to be the most efficient and economical for this purpose, while his arrangement of the elements in series is still found to be the arrangement which produces the highest electric potential. The use of zinc in battery construction has never been superseded. It has been employed in nearly every battery that has ever been invented, and enters into the construction of every one now in general use. And copper, in connecction with it, is the next metal in most general use for this purpose.

Cell, Element, and Battery.—A single pair of metals or their equivalent, with the fluid and containing vessel, or substance, is designated as a *cell* or *element*, and a combination of such cells is called a *battery ;* the latter term being also applied to a single cell, when employed alone.

Battery Sign.—This sign, |ı|ı|ı, is used to represent the battery; the short, heavy lines representing the zinc, and the light ones the copper or its equivalent; the number of lines varying indefinitely, according to the size of the battery, each pair representing a cell.

Electrodes and Poles.—Since the metals, or their equivalents, are the principal avenues in which the electricity travels, downward through the zinc and upward through the copper, or its equivalent, they are called the *electrodes*, from ἠλεκτρον ὁδος, electric road. The zinc, being consumed by the chemical reaction, is termed the *soluble* or *generating* electrode, and the copper the *conducting* electrode. This term is also applied to instruments used for conveying and applying electricity.

The parts of the electrodes which project out of the fluid are known as the *poles;* the projecting part of the zinc being designated as the *negative* pole, and that of the copper, or its equivalent, as the *positive* pole. These terms are also applied respectively to the outer terminals of the conducting wires connected with the poles of a battery or other electric generator.

The terms *positive* and *negative* are also applied to the electrodes, the zinc being called the *negative* electrode, and the copper, or its equivalent, the *positive*.

Conditions of Electric Energy.—In estimating the electric energy of a cell three important conditions are to be considered, termed respectively *electromotive force*, *resistance*, and *current;* any two of which being known, the third can be ascertained by calculation.

Electromotive Force.—Electromotive force, symbol E. M. F., has been defined as "that which moves or tends to move electricity from one point to another." It is represented by difference of electric potential; electricity always moving, or tending to move, from higher to lower potential with a force, or pressure, equal to this difference. This condition, in the cell, depends on the nature of the materials employed and their mutual relations, varying in proportion to the chemical reaction between the soluble electrode and fluid, and the resistance to such reaction and the electric, molecular movement generated by it, by the various materials composing the cell. Hence this is not properly a force, but a condition producing force.

Resistance.—Resistance, symbol R, is that which opposes the movement of electricity through a conductor; and, in the cell, it depends chiefly on the nature of the fluid, the quantity intervening between the electrodes, and a certain effect termed polarization. It varies directly as the length and inversely as the cross-section of the conductor; and since the distance between the electrodes may be regarded as the length of the fluid conductor, while the area of their immersed surfaces measures its cross-section, the fluid resistance of a cell varies directly as the distance between the electrodes, and iuversely as the area of their immersed surfaces; hence the least resistance, dependent on these conditions, is obtained with the shortest practicable distance between the electrodes coupled with the greatest area of immersed surface.

The resistance of battery fluids varies greatly; that of pure water or acid alone, for instance, is very high, but in mixtures of the two the resistance is greatly reduced. Hence the importance of selecting the fluid with reference to its resistance as well as its chemical reaction.

Current.—Current, symbol C, is the electric movement produced in a conductor by electromotive force in opposition to resistance; its value being ascertained by dividing the former by the latter. Hence strength of current varies as each of these factors, increasing with increase of E. M. F. or decrease of resistance, and decreasing with decrease of E. M. F. or increase of resistance, but remaining constant when each varies in the same ratio as the other,

Units of Electromotive Force, Resistance and Current.— The Volt is the unit of electromotive force, represented practically by the E. M. F. of the Daniell cell, to be described hereafter, to which it is nearly equal.

The Ohm is the unit of electric resistance, represented by the electric resistance of a column of mercury 106 centimeters in vertical height, and 1 square millimeter in cross-section, at the temperature of 0° C.

The Ampere is the unit of current strength, represented by an E. M. F. of 1 volt divided by a resistance of 1 ohm.

As electric measurement pertains to a future chapter in which it is resumed and treated at greater length, the above brief definitions of the three principal electric units must suffice for our present purpose.

Operation of the Voltaic Cell.—If the metals are strictly pure there is no perceptible action either chemical or electric in the voltaic cell so long as there is no connection between the electrodes; but when the poles are brought into contact, or connected by a conductor, chemical reaction, accompanied by the generation of electricity, begins at once. If the metals are impure, as is usually the case, chemical reaction and electric generation, in a limited degree, occur without polar connection. In either case the water is decomposed, the hydrogen collecting on the surface of the copper, and the oxygen combining with the zinc, forming oxide of zinc, which then combines with the sulphuric acid, forming sulphate of zinc. The generation of electricity may be proved by separating the poles slightly, when an electric spark will pass between them.

Theory of Electric Generation in the Cell.—Volta, as we have seen, attributed the electric generation to the contact of the metals, and this was the accepted theory among scientific observers to the time of Faraday.

THE VOLTAIC BATTERY. DEFINITIONS.

Meantime chemistry, almost unknown as a science in Volta's time, had made rapid advancement, and Faraday's observations having led him to the conclusion that the mere contact of the metals was not an adequate cause for the results obtained, and was not proportionate to such results, made an investigation of the relations between the chemical and electric actions of the cell, which enabled him to demonstrate that the electric generation was in exact proportion to the chemical reaction; and his results having been fully verified by other observers, the chemical theory of electric generation in the cell has since been generally accepted as correct. It may be briefly stated as follows:

The principal seat of chemical reaction is at the surface of the zinc, which is consumed by oxidation, while the copper acts as a conductor and is not consumed. Hence, since electric movement is from higher to lower potential, and the same law applies to the energy of chemical reaction, in common with other forms of physical energy, and since the electric energy of the cell is found to be strictly proportionate to its chemical reaction, it is assumed that the electric current originates at the surface of the zinc and flows through the fluid to the copper.

In the absence of external connection between the metals, it is evident that the difference of electric potential would immediately become equalized and the current cease, but when they are brought into external contact, or connected by a conductor, the current finds an outlet through the copper, and flows back to the zinc through the external circuit; chemical reaction is thus sustained and the current becomes continuous.

The electric generation produced by *Zamboni's dry pile* is adduced in proof of the contact theory. This pile was made of a large number of paper disks, some thousands, coated with zinc or tin foil on one surface, and with dioxide of manganese on the other, and closely compressed in a glass tube, their similarly coated surfaces turned in the same direction, bringing those oppositely coated into contact. Such a pile, when its circuit is completed, as in Volta's pile, can excite the electroscope, ring a bell, or give sparks. But this electric action can be accounted for by chemical reaction, caused by dampness in the paper, rather than by the mere contact of different substances.

Such experiments as the divergence of the leaves of the electroscope and the oscillations of the magnetic needle by the mere contact of different metals in their immediate vicinity are also adduced in support of the contact theory; but such electric action is doubtless due to the static charge generated by the slight friction produced in making the contacts.

The law of the conservation of energy requires the expenditure of energy in one form as a condition of the production of the same amount in another form. Now in every electric generator, static or dynamic, machine or battery, this law is found to be strictly true; there must be a complete circuit of materials differing in molecular constitution, and the expenditure of energy, mechanical, chemical, or in some other form, at some point in the circuit as a condition of electric generation; and this expenditure must be equal in amount to the electric energy produced and that absorbed by friction, heat, or otherwise. Hence as chemical energy is the only energy expended in the battery, the conclusion is inevitable that it is the source of the electric energy generated.

Amalgamation of the Zinc.—As strictly pure zinc is too expensive for practical use in battery cells, and ordinary commercial zinc contains a certain percentage of iron and other metals by which chemical and electric action is generated independent of the copper, and the energy thus, in part, expended within the cell, without passing through the external part of the circuit, where it can be made available, a fault known as *local action*, the method has been adopted of amalgamating the surface of the zinc with mercury, which renders it more homogeneous and prevents any serious interference from local action, which is thus reduced to its minimum.

The zinc is first cleansed with potash or otherwise, after which the mercury, mixed with acid, is applied by any convenient method, or the zinc dipped into the mixture. Sulphuric acid may be used for this purpose, but a mixture of five parts chlorhydric and one part nitric acid is preferable. The same result is also obtained by adding bisulphate of mercury to the solution, the mercury combining with the zinc, and the acid being set free. Amalgamation is thus more easily accomplished and better sustained.

The molten zinc, before it is cast into plates, may be permanently amalgamated by the addition of about 4 per cent of mercury, and thus the frequent renewal, necessary with surface amalgamation, be dispensed with.

Insulation and Clamping.—When both electrodes are suspended from the support, they must be insulated from each other, either by making the support of insulating material, or insulating one of them from it. They must also be provided with clamps and binding-screws for making connections, and the points of contact with conductors kept clean and free from oxidation.

Polarization.—It has been stated that, as a result of the chemical reaction of the cell, hydrogen accumulates on the surface of the copper. As this accumulation increases, it weakens the electric action and finally stops it; an effect termed *polarization*. As a thorough knowledge of this effect and the methods used to correct it is of the highest importance in the study of the cell, it is proper first to examine its nature.

If the poles of a battery of two or more cells be connected with platinum terminals which project into a vessel of water acidulated with sulphuric acid, hydrogen will be evolved at the terminal connected with the zinc, or negative pole, and oxygen at that connected with the copper, or positive pole, in the exact proportions which form water, two volumes of hydrogen and one of oxygen. If now the battery be disconnected, and the terminals of the wires connected with the gas tubes brought into contact, an electric current will flow through them in the reverse order to that of the original current, the gases, at the same time, recombining to form water. From which it is evident that the electric energy expended in decomposing the water was stored up in the gases, and reappears when they return to their original state.

Fig. 1 shows the apparatus by which this decomposition is effected; oxygen being evolved in the right-hand tube and hydrogen in the left. It will be noticed that the hydrogen is evolved at the pole *towards* which the current flows within the decomposing vessel, connected externally with the zinc of the battery, and the oxygen at the pole *from* which it flows, connected externally with the copper of the battery; the external current through the wires connecting with the battery being *towards* the oxygen tube and *from* the hydrogen tube; also that the same direction of current-flow occurs with respect to the battery, internally from zinc to copper, externally from copper to zinc, completing the circuit through the decomposing vessel. And since the current from the gas tubes, when disconnected from the battery and brought into mutual contact, flows in the reverse order to that of the original battery current, it is evident that when the gases accumulate on the electrodes within the cell, the effect must be to set up a similar re-

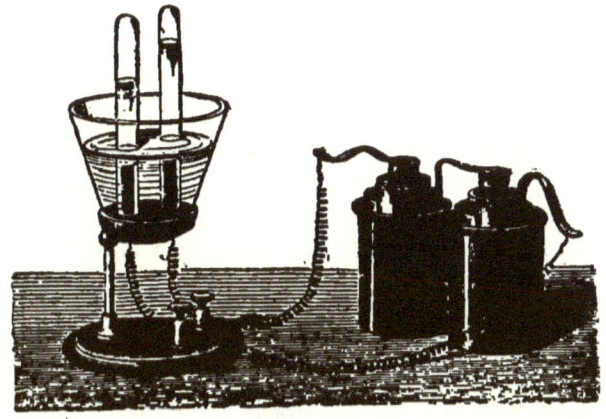

FIG. 1.

verse current, which neutralizes the primary current. Hence this action is appropriately termed *polarization*, since it produces opposing poles.

But since the oxygen, from its strong affinity for the base metals, combines with the zinc, the polarization is confined to the hydrogen, taking place on the copper. This affinity of the oxygen makes the use of a platinum terminal necessary for the oxygen at least, when it is desired to collect the gases separately, as above, since oxygen does not combine with platinum; while, if a base metal were used, it would become oxidized, and no oxygen gas could be collected.

A single cell of less E. M. F. than $1.49\frac{1}{2}$ volts is insufficient to decompose water, since the polarizing energy, in such case, exceeds the generating energy; hence two such cells at least are required.

To correct polarization the accumulation of the hydrogen must be suppressed, and to do this in the most effectual, practical, and economical way, without impairing the energy of the cell in other respects, is the most important problem in cell construction. It may be done either by mechanical or chemical means, the latter being the most practical and effectual. Among the mechanical means adopted are the lifting of the electrodes, or the conducting electrode alone, out of the fluid, so that the hydrogen may pass off. With a battery of two or more cells the electrodes of half the cells may thus be depolarized while the other half remain in the fluid and furnish the current. And as only a momentary exposure is required, any simple mechanism, operated by a weight or spring, by which this alternate exposure can be effected will answer the purpose. Another method is the injection of air into the fluid against the conducting electrode. Either of these methods may be employed for work which does not require a continuous, strong current; and they are sometimes used in connection with the chemical process to intensify the electric action. But all mechanical contrivances for this purpose are necessarily cumbersome and inconvenient, and hence undesirable.

The chemical method is to introduce into the cell some substance which has a strong chemical affinity for the hydrogen, and absorbs it without interfering with the action of the cell in other respects. This is accomplished either by the use of a single fluid holding the substance in solution, or by using two fluids separated by a porous cup or otherwise, so that the zinc shall be in contact with one fluid, and the copper, or its equivalent, in contact with the other. Hence arises the division of cells into two classes, *one*-fluid and *two*-fluid cells, each of which now claims our attention.

CHAPTER II.

ONE-FLUID CELLS.

Smee's Cell.—This cell, represented by Fig 2, was invented by Smee, an English electrician, in 1840. The electrodes consisted originally of a plate of platinum suspended between two plates of zinc; the object of this arrangement being to utilize both surfaces of the platinum, since, in any cell, only the adjacent surfaces of the opposite electrodes are brought into action. Depolarization was effected by platinizing the surface of the platinum, electrically, so as to furnish a rough surface from which the hydrogen could escape much more freely than from a smooth surface, since a point has neither adhesion nor electric resistance, and the hydrogen atoms, being at the same electric potential, are self-repellent.

FIG. 2.

The platinum plate was subsequently replaced by a platinized silver plate, and this was afterwards replaced by a copper plate, covered with a rough coating of copper, then silver-plated and then platinized. The fluid consists of one part sulphuric acid to seven parts water.

This cell is practical and efficient, where constancy of current is not required; but, like all single-fluid cells, the current soon weakens.

Zinc-Carbon Cells.—The expense of constructing cells with platinum or silver stimulated the search for some cheaper material, and Sir William Grove first suggested the use of carbon, but failed to reduce his suggestion to practice. In 1843 Bunsen constructed the first cell in which carbon was used. This was a two-fluid cell, and will be described under that head. Since that time carbon has been successfully employed in the construction of numerous different cells which have come into general use.

Carbon suitable for this purpose may be obtained from the inside of gas-retorts, and cut into plates or other convenient forms. It may also be prepared from coal, coke, graphite, or charcoal, pulverized, cemented together, reduced to the proper form in moulds; then dried, baked, and soaked in sirup of sugar repeatedly, till it acquires the requisite density and firmness.

To obtain a good connection for the clamps and conducting wires, it is desirable that the upper part of the carbon should be soaked in melted paraffine, and then copper-plated. The paraffine fills the pores, and excludes the acid, which would otherwise ascend by capillary attraction and destroy the copper.

The advantages of carbon are: 1. That it is cheap. 2. That, like platinum, it is insoluble in acid, and possesses the conductivity necessary for an electrode. 3. That it has a rough surface, similar to that produced artificially with platinum in Smee's cell, by which depolarization is assisted. 4. That, being porous, a great amount of internal surface is brought into contact with the fluid.

Walker's Cell.—Walker was one of the first to use carbon as an electrode. In 1849 he constructed a cell similar to the Voltaic, substituting carbon for copper. In 1857 he platinized the carbon, copper-plated and tinned its upper end, placed the lower end of the zinc in a vessel of mercury, by which it was kept amalgamated; and used a fluid composed of one part, by volume, of sulphuric acid to eight parts of water.

This cell has great constancy, requires but little care, and is cheaply constructed. Its electric energy is about the same as that of the Smee cell. It has been extensively used in England for telegraphing, with great success.

Potassium Bichromate Cell.—The most efficient single-fluid, carbon and zinc cell is that in which potassium bichromate is the depolarizing agent. The fluid consists of water, sulphuric acid, and potassium bichromate, and the following are recommended as the best proportions:

66 per cent by weight of water,
25 " " " " " sulphuric acid,
9 " " " " " potassium bichromate.

The bichromate is decomposed by the sulphuric acid, and oxygen liberated, which enters into combination with the hydrogen while both are in the nascent state, producing water, and thus preventing the accumulation of the hydrogen. Practically, however, there is a certain amount of polarization, and salts are also formed, which, if allowed to accumulate, reduce the conductivity of the carbon; so that the intensity of the electric action soon diminishes, and the fluid requires to be agitated, either by injecting air into it, or by withdrawing the electrodes, or the zinc alone. Air may be injected through a rubber tube; but this method, though very effective, is inconvenient in practice, and the withdrawal of the electrodes is preferable. Hence this cell is best adapted to work where constancy is not required; so that after a few minutes' use the electrodes may be withdrawn and the cell allowed to recuperate, while preparation is made for the next operation. Medical, surgical, and laboratory work is of this character, and for such work it is especially fitted; having the highest electric energy of any single-fluid cell in use; being capable of application to a great variety of different operations; being free from noxious fumes; and easily made portable, either as a single cell, or a battery of cells.

It is usually fitted with a hard-rubber cover, to which the electrodes are attached, and thus insulated. And as depolarization is in proportion to the relative amount of surface of the conducting electrode brought into action as compared with that of the soluble electrode, it is usual to have a carbon plate on each side of the zinc plate; using two carbons and one zinc, or three carbons and two zincs.

As this fluid soon weakens with use, and is subject to slow chemical change when not in use, the amount should be so proportioned to the size of the electrodes as to prevent rapid exhaustion.

The Grenet Cell.—The bottle form of the bichromate cell, known as the Grenet, shown in Fig. 3, is convenient for work requiring only a single cell. The electrodes are attached to a close-fitting hard-rubber cover, and the zinc is connected with a sliding rod by which it can be drawn up into the wide neck, while the enlarged base gives the requisite capacity for a full supply of fluid.

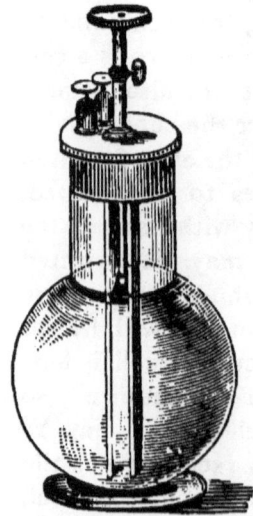

FIG. 3.

The zinc of any bichromate cell should be kept well amalgamated, and when the fluid is renewed, the deposit of chrome alum which accumulates in the bottom of the vessel should be removed, and the carbons soaked in warm water to remove similar deposits from their pores.

ONE-FLUID CELLS.

The Mercuric Bisulphate Cell.—This cell is extensively used for medical pocket-batteries, which are usually constructed with two small zinc and carbon cells, each about an inch square and half an inch deep. The carbon is placed in the bottom of a hard-rubber cup, and the zinc, resting on a ledge which insulates it, forms the cover.

The fluid consists of a solution of mercuric bisulphate in water; a few grains of the bisulphate to a teaspoonful of water being sufficient for a cell. The acid of the bisulphate unites with the zinc, setting the mercury free, which keeps the zinc amalgamated.

The solution can be made up quickly, and renewed when wanted; and the cell is easily cleaned and requires but little care.

The Leclanché Cell.—Leclanché, a French electrician, was the first to use sal-ammoniac (NH_4Cl) in the construction of battery cells. Fig. 4 represents this cell, which consists of a glass jar, in the centre of which is placed a porous cup containing a carbon plate, which projects above it as shown, and is surrounded with crushed carbon and crystals of manganese binoxide, mixed in about equal proportions. This cup is closed with Portland cement, except two small openings left for ventilation, and its contents constitute the conducting electrode. The zinc is a round

FIG. 4.

rod, about half an inch in diameter, placed in a recess provided for it in the outer vessel.

The fluid is a saturated solution of sal-ammoniac in water; about 6 oz. of the salt being required for a quart cell, which is kept about two thirds full, and its upper surface coated with paraffine, to prevent surface accumulation of the salt. This solution permeates the porous cup and materials contained in it, a little water being added through the ventilating openings.

The manganese binoxide being rich in oxygen, which is evolved by the chemical action, acts as a depolarizer, the oxygen uniting with the hydrogen to form water; the large proportion of surface in this electrode to that of the zinc greatly facilitating depolarization. But if electric action is continued too long at a time, an excess of hydrogen accumulates, oxygen not being generated with sufficient rapidity to unite with it, and polarization ensues, requiring a period of rest for the absorption of the hydrogen. Hence it is not fitted for work on a continuously closed circuit.

The E. M. F. of this cell is about 1.48 volts, and its resistance comparatively low, being reduced by the improved conductivity of the mixture constituting the conducting electrode. The current is always in full proportion to the consumption of material, there being no chemical action except with a closed external circuit, and hence no waste of material by local action or otherwise. The electrodes can therefore remain permanently immersed in the fluid without detriment, so that the cell can remain undisturbed till the fluid is exhausted, a little water being added occasionally to supply the loss by evaporation. It contains no poisonous materials, emits no noxious fumes, and can endure a temperature of $-16°$ C. without freezing or decrease of electric energy.

ONE-FLUID CELLS.

The above style of Leclanché cell is known as the *Disque*. In a more recent style, known as the *Prism* or *Gonda*, the porous cup is dispensed with, and the conducting electrode constructed with two prisms, attached by stout rubber bands to the central carbon plate as shown in Fig. 5; spaces for the circulation of the fluid being left between the plate and prisms.

These prisms are composed of the double chloride of iron and ammonia, mixed either with manganese binoxide, graphite, or powdered retort-carbon, as preferred, and cemented together with any suitable glutinous substance, as tar, rosin, or gum-lac. The greater compactness of this form of electrode gives it higher conductivity than the Disque form, while the suppression of the porous cup reduces the resistance. The electrodes are suspended from a close-fitting cover of insulating material by enlarged pole-pieces, which close the openings through which they pass.

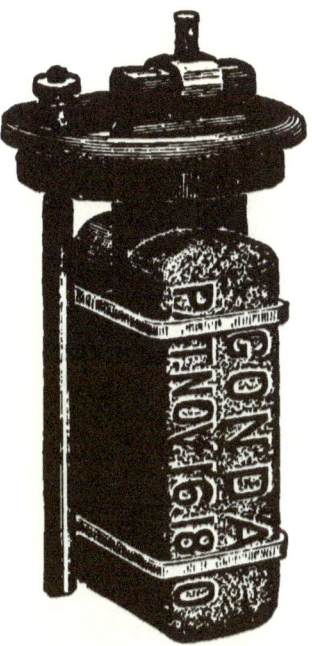

FIG. 5.

The Leclanché cell, in both styles, has come into extensive use for *open circuit* work in which there are continually recurring intervals of rest; and in France it is used for telegraphing, to which it is found to be well adapted in offices where the work required is not so constant as to cause inconvenience from polarization; cells having been used for nine years without renewal of the zincs, and only one renewal of the sal-ammoniac.

The success of the Leclanché has given rise to a number of similar cells; carbon and manganese binoxide, variously combined, being employed as the conducting electrode; the soluble electrode in all of them being a zinc rod, as in the Lelanché.

The Law Cell.—Prominent among these is the Law cell, the electrodes of which are shown in Fig. 6. The conducting electrode consists of two hollow cylinders, one inclosed within the other, with space between them. The zinc is placed in a vertical opening in the same side of both carbons, through which the fluid can circulate freely, and is thus brought into closer proximity to the conducting electrode than in the Leclanché, and the fluid resistance thereby reduced and depolarization made more rapid and effective. Depolarization is also made more effective by the increased proportion of surface in the conducting electrode to that in the zinc. Both electrodes are attached to a close-fitting, insulating cover.

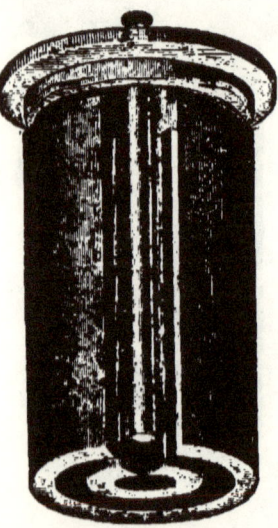

Fig. 6.

The Diamond Carbon Cell.—The Diamond Carbon Cell, shown in Fig. 7, is another of the same class, in which the conducting electrode consists of seven round rods arranged in a circle around the zinc rod, and put in electric connection with each other by attachment to a cover made of the alloy known as white metal, which is not easily oxidized; the zinc being insulated by a porcelain bushing.

This cell has the same advantages in regard to de-

ONE-FLUID CELLS. 21

polarization as the last. The metal cover reduces its resistance, and the separate rods are easily renovated by heating or soaking in hot water when necessary, and cheaply replaced when worn out.

In other cells of this class, as the Laclede and Mi-

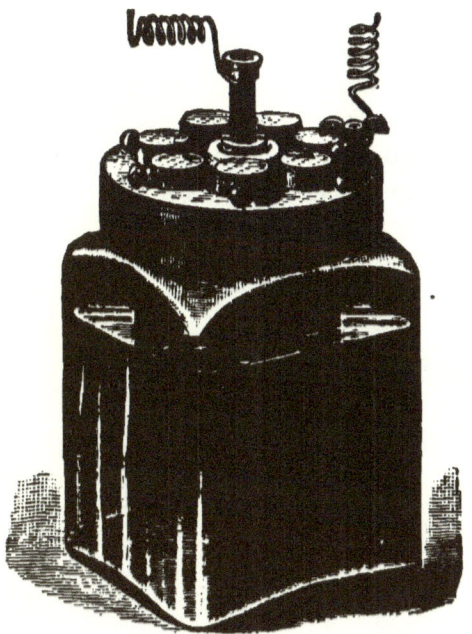

FIG. 7.

crophone, the cover is made a part of the conducting electrode and the zinc insulated from it.

The ringing of electric bells is one of the most common uses to which sal-ammoniac cells are applied, and for which their constancy on open-circuit work especially fits them.

Dry Cells.—A cell constructed with a semi-fluid, not liable to spill, is termed *dry*. Cells filled with sand or sawdust, soaked with dilute acid, are instances of this construction. Portable cells, having starch or similar material to absorb the fluid, and hermetically sealed, are now becoming common, and are very convenient for many purposes ; and, when properly constructed, have a high degree of constancy and efficiency. An absolutely dry cell is an impossibility ; a certain degree of dampness or moisture being essential to proper chemical action.

Polarization of One-Fluid Cells.—All one-fluid cells, no matter how perfect their construction, are subject to polarization to a greater or less degree ; and though less complicated than two-fluid cells, and more convenient for many uses, they are not adapted to work requiring a continuous current, or in which the intervals of rest are not sufficient for complete depolarization.

CHAPTER III.

TWO-FLUID CELLS. BATTERY FORMATION.

Construction of Two-Fluid Cells.—In the two-fluid cell polarization is either wholly prevented, or so reduced that the cell may be used for work requiring greater constancy than can be obtained from a one-fluid cell. The construction requires that the conducting electrode shall be surrounded with a fluid capable of suppressing the hydrogen, while the soluble electrode is surrounded with a fluid capable of chemical combination with the material of which the electrode is composed; and that the means of separation between the fluids shall not be such as to prevent electric or chemical action. For this purpose a porous cup, like that in the Leclanché cell, made of unglazed porcelain, is placed inside the larger vessel, and contains one of the electrodes with its fluid, while the other electrode with its fluid is placed in the outer vessel, and electric and chemical action takes place through the pores of this cup, where the fluids come into contact.

Various other means of separating the fluids are used, as vessels or partitions of wood, paper, or animal membrane. Gravitation is also employed; a heavy fluid being used in connection with a light fluid, the former settling to the bottom of the vessel, while the latter rises above it.

The Daniell Cell.—This is one of the oldest and best two-fluid cells in use. In was invented by Daniell, an English electrician, in 1836, and has undergone various modifications. Fig. 8 represents one of the best known styles. The outer vessel is a glass jar containing water

24 DYNAMIC ELECTRICITY AND MAGNETISM.

or dilute sulphuric acid, in which is placed a hollow cylinder of zinc, having a slit in one side for the free circulation of the fluid. Inside this cylinder is placed a porous cup containing a solution of copper sulphate in water, to which some crystals of the sulphate are

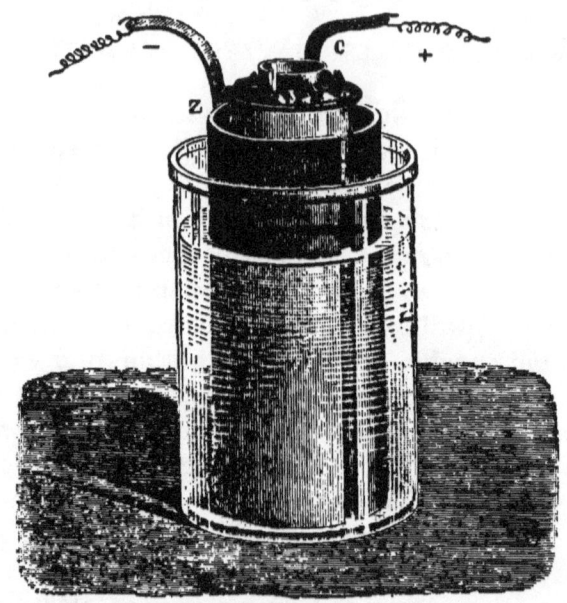

Fig. 8.

added; in which is placed a copper cylinder, slit like the zinc.

The chemical reaction is as follows: Hydrogen being liberated by the oxidation of the zinc, and the copper sulphate ($CuSO_4$) decomposed, the copper (Cu) is deposited on the copper cylinder, and the other constituent (SO_4) unites with the hydrogen (H_2), forming sulphuric acid (H_2SO_4), which in turn is decomposed by the zinc (Zn), forming zinc sulphate ($ZnSO_4$), more hydrogen being set free to unite with the liberated SO_4, as before; the interchange taking place through the pores of the inner vessel. The hydrogen being thus entirely suppressed, depolarization is complete, and the copper cylinder, accumulating only pure copper, is always in the best condition as an electrode.

The E. M. F. of this cell is about 1.05 volts. It has great constancy, and is but slightly affected by changes of temperature ranging from $+18°$ to $+100°$ C.; below this range the internal resistance increases, and at $-5°$ to $-7°$ C. the solution freezes. Its chief defect is that the consumption of material is nearly as great when unemployed as when employed. Amalgamation of the zinc is not necessary.

The electric resistance of the porous cup, which results from the reduction of the cross-section of the fluid in passing through the pores, and from local action caused by the material of which this cup is composed, led to the invention of cells in which the fluids are separated by gravity.

The Callaud Cell.—One of the best known gravity cells is the Callaud, represented by Fig. 9. The copper is placed in a solution of copper sulphate at the bottom of the vessel, and the zinc suspended in a solution of zinc sulphate near the top; the two fluids being kept separate by the difference in their specific gravity, the copper sulphate being the heavier. Connection with the copper is made by a copper wire, insulated by gutta-percha or India-rubber to protect it from injury by local action at the junction of the fluids, and from contact with the zinc.

FIG. 9.

The separation of the fluids is never quite complete; a certain percentage of the copper sulphate rising to the upper part of the vessel, producing a copper deposit on the zinc; an effect which is increased by local action on both electrodes, evolving hydrogen and producing ascending and descending currents. As this deposit accumulates copper pendants are formed, which increase in length till they reach the copper sulphate, when this action becomes much more rapid, with increased waste of the copper sulphate. Hence they should be removed before attaining this length, by lightly tapping the zinc, causing them to drop off.

This cell has about the same E. M. F. as the Daniell, while the reduction of resistance by the removal of the porous cup produces a corresponding increase of current in the external circuit.

The Grove Cell.—This cell was invented by Grove, an English electrician, in 1839. It is constructed with an amalgamated zinc cylinder immersed in dilute sulphuric acid, contained in a glass jar, within which is a porous cup containing a strip of platinum immersed in strong nitric acid. This acid is rich in oxygen, which unites with the hydrogen, producing complete depolarization; and, being a good electric conductor, greatly reduces the resistance. The chemical reaction forms water, and also nitric tetroxide (N_2O_4), which is emitted in noxious, red fumes, and is one of the greatest objections to this otherwise excellent cell.

Its E. M. F. is about 1.8 volts, which is 80 per cent greater than that of the Daniell cell, while its internal resistance is about 20 per cent that of the Daniell. Hence a Grove cell of the same size as a Daniell has about nine times the current strength. It is therefore one of the most powerful cells in use.

The Bunsen Cell.—In 1843, Bunsen, a German electrician, adopting the plan originally proposed by Grove, produced a cell having carbon instead of platinum in the porous cup, but otherwise identical with the Grove cell. This substitution greatly reduced the cost without impairing the energy; the E. M. F. and internal resistance being about the same as in the Grove.

Depolarization is complete, but the same noxious fumes occur as in the Grove cell.

The Silver Chloride Cell.—Silver chloride was first used in the construction of battery cells by Marié Davy about 1860; subsequently Warren De La Rue made such improvements in the construction as to bring the cell into general use. His cell, as shown in Fig. 10,

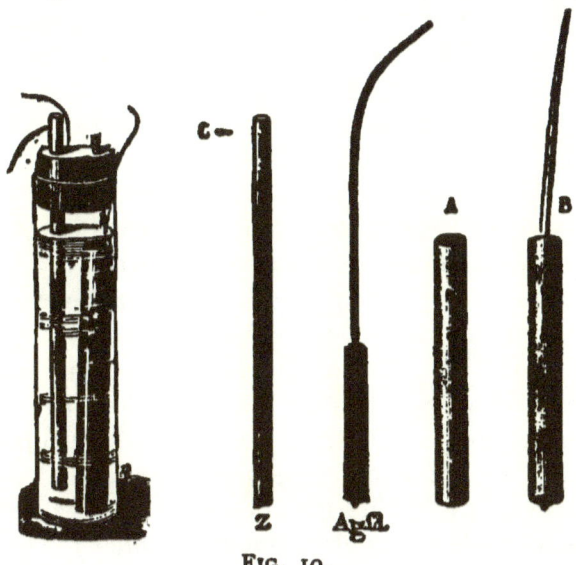

FIG. 10.

consists of a small glass jar, about 5 inches in height and 1½ inches in diameter, which contains a dilute solution of sal-ammoniac, in the proportion of 23 grammes to 1 liter of distilled water, in which is placed a small rod of unamalgamated zinc of superior quality; also a strip of silver imbedded in a small cylinder of silver chloride (AgCl), which is contained in a cylinder of parchment-paper. The electrodes are shown separately,

Z representing the zinc, $Ag.Cl$. the imbedded silver strip, A the paper cylinder, and B the cylinder and inclosed strip. The cell is closed by a paraffine stopper fitting air-tight, through which the electrodes protrude. This prevents evaporation and creeping salts, and insulates the electrodes from each other. Near the top of the zinc there is a hole into which the silver strip, bent over from the adjoining cell, enters as shown at C, when the cells are connected into a battery.

The silver chloride in this cell acts as a depolarizer much in the same manner as the copper sulphate in the Daniell cell. By its decomposition zinc chloride is formed and silver deposited on the conducting electrode; hence there is no oxidation, no deposit of hydrogen, and consequently no polarization.

The E. M. F. is 1.03 volts, and the internal resistance 4.3 ohms. Resistance, being chiefly due to the silver chloride, is much greater when the cell is first used than subsequently when reduced by the deposit of silver throughout the mass of the chloride.

The small size of this cell makes it convenient for the construction of batteries having a large number of cells, one constructed by De La Rue containing 11,000.

The construction of battery cells is limited only by the number of combinations of suitable materials which may be formed; and as the principles which govern these combinations have been fully set forth and illustrated by the various cells described in the preceding pages, it is unnecessary to carry these details farther.

Battery Formation.—There are two principal methods of combining cells to form batteries, known by the terms *series* and *parallel*. When joined in series, the soluble electrode of each cell is connected with the conducting electrode of the adjoining cell; and when joined in parallel, all the soluble electrodes are connected with each

TWO-FLUID CELLS. BATTERY FORMATION.

other, and likewise all the conducting electrodes. The latter method is also known by the terms *multiple arc, side by side,* and *for quantity;* and the series method by the term *for intensity,* to distinguish it from the method *for quantity.* But as the use of these various terms is confusing and unnecessary, it is better to confine ourselves to the terms first given above, which have received the sanction of leading electricians and are now in general use.

With a given number of cells a given amount of electric energy may be generated, which it is evidently impossible, according to the law of the conservation of energy, either to increase or diminish by any method of connecting them.

But it is possible to control and direct this energy in such a manner as shall best subserve the uses to which it is to be applied; and, for this purpose, either the series method or the parallel may be used alone, or the two combined to any desired extent.

Fig. 11 illustrates the method by which six cells may be joined in series; the circles representing cells, and the lines conductors.

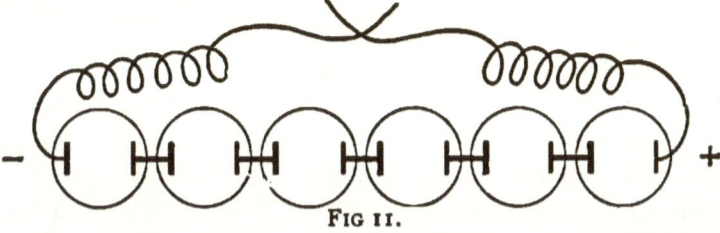

FIG 11.

Fig. 12 shows how the same cells may be joined in parallel.

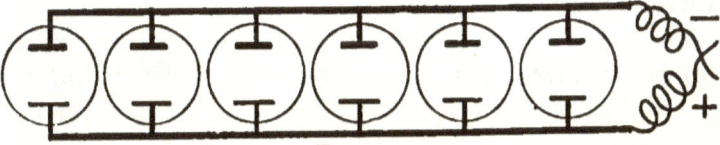

FIG. 12.

30 DYNAMIC ELECTRICITY AND MAGNETISM.

Fig. 13 shows a combination of two series of three cells each, and these series joined in parallel; and Fig.

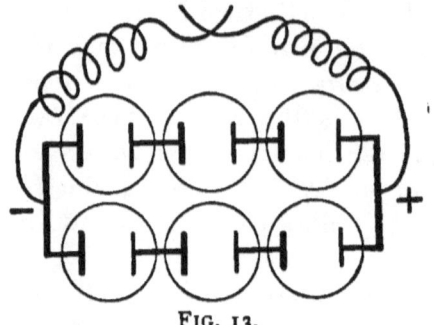

FIG. 13.

14 shows three series of two cells each, and these three joined in parallel.

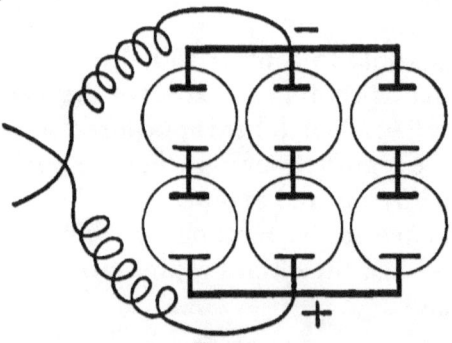

FIG. 14.

Since electromotive force is that which moves or tends to move electricity from one point to another, and depends on difference of potential, and since this difference, in a cell, depends on the nature of its materials and the method of construction, we should expect to find the E. M. F. of a small cell equal to that of a larger one of the same composition and construction; and experiment proves that such is the fact.

The case is analogous to that in hydrostatics, where liquid pressure depends on difference of level and not on the size of the vessel; the liquid in a small vertical

TWO-FLUID CELLS. BATTERY FORMATION. 31

tube balancing liquid of the same kind contained in a larger one connected with it, so that the level is the same in each. So when a small cell is joined to a larger one of the same kind by connecting the similar electrodes, so that opposing currents meet, the current from the one exactly neutralizes that from the other; proving that both currents have the same strength, and hence that the E. M. F. of each cell is the same. But this is not true of dissimilar cells, differing by construction in E. M. F., nor similarly in hydrostatics of liquids differing in specific gravity.

The six cells in Fig. 12, joined in parallel, are practically equivalent to one cell six times the size of any one of them; for, the similar electrodes of each kind being joined together, each set acts as one electrode. Hence the E. M. F. of the battery, connected in this way, is only equal to that of a single cell; just as in hydrostatics the liquid pressure in six tubes of the same size placed vertically side by side, and connected with a horizontal tube at bottom, is only the same as that in any one of the tubes alone. But if the six are joined end to end in a vertical series, the pressure becomes six times as great. So if the six cells are joined in series, as in Fig. 11, the E. M. F. becomes six times as great. In the former instance we have *liquid* pressure, in the latter *electric* pressure. But since electric resistance varies directly as the length and inversely as the cross-section of a conductor, the resistance also becomes six times as great; each of the six cells with its electrodes and fluid adding a unit to the length of the conducting line, while the cross-section remains the same. And since the *quantity* of electricity passing through a conductor, represented by the volume of current, equals the E. M. F. divided by the resistance, the quantity obtained from the series in Fig. 11 is only one sixth of that obtained from the six cells in parallel, as in Fig. 12, and hence no greater than that of a single cell, though the E. M. F. is six times as great.

When the six cells are joined in parallel, as in Fig. 12, the resistance is only one sixth of that developed in a single cell; for the cross-section of the united conductors is six times as great, affording six times as large an avenue for the passage of electricity. Hence, though the E. M. F. is only equal to that of a single cell, the electric quantity or volume of current is six times as great, and hence also six times as great as that of the six cells in series, which has been shown to be only equal to that of a single cell.

Hence in the series combination we have current *intensity* at the expense of current *quantity*, small current and large E. M. F.; and in the parallel combination, quantity at the expense of intensity, large current and small E. M. F.; one being in the inverse ratio of the other in each case.

The combination proper to be used depends on the nature of the required work. If there is high resistance to be overcome, as in a long telegraph line, the intensity must be sufficient to overcome it, and leave a sufficient surplus to operate the instruments, and the series arrangement should have the preference. But if the resistance is low, and the required quantity large, as in the deposition of metal in electro-plating, the parallel arrangement is to be preferred.

The practical rule is to *make the internal resistance of the battery equal to the external resistance to be overcome:* and our illustrations show that the variation of the relative proportions of quantity and intensity by different methods of combination is practical for this purpose to any required extent.

The correctness of the rule becomes evident when we consider that if by a preponderance of the series arrangement the internal resistance exceeds the external,

TWO-FLUID CELLS. BATTERY FORMATION. 33

greater intensity is developed than is required; and if by a preponderance of the parallel arrangement the internal resistance is less than the external, the intensity is insufficient to overcome the external resistance. In the former case there is a waste of intensity at the expense of quantity, and in the latter a waste of quantity at the expense of intensity. So that the most economical arrangement is attained by following the rule given above, which is based on the intensity or quantity required, to which equality of internal and external resistance serves merely as a convenient guide.

In all the various combinations of a given number of cells which may be made, as shown, the product obtained by multiplication of the current strength in amperes into the E. M. F. in volts must remain the same, since each factor varies inversely as the other; hence difference of combination can produce no variation in the amount of electric energy developed, as represented by this product, the variation observed pertaining exclusively to the different factors.

Connection between Cells.—Since all unnecessary resistance within the battery causes a waste of energy, it is important that this resistance should be reduced to the minimum, both in the cell itself, as we have already seen, and in the connection between the cells. This can often be accomplished by clamping the electrodes of adjoining cells together without any intervening connection. Fig. 15 shows a form of the Grove cell specially adapted to this purpose. The cell being rectangular, the porous cup thin and flat, and the electrodes flat strips, permits a compact arrange-

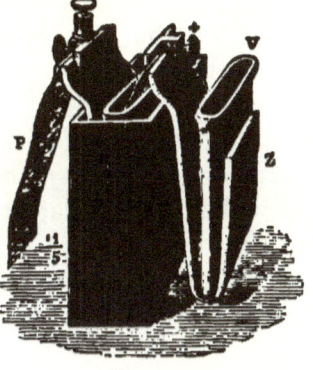

FIG. 15.

ment of all the parts; the zinc, Z, being bent so as to inclose the porous cup, V, while its upper end is clamped in immediate contact with the platinum, P.

Fig. 16 shows the silver chloride battery, in which the cells are round but small, permitting them to be placed so

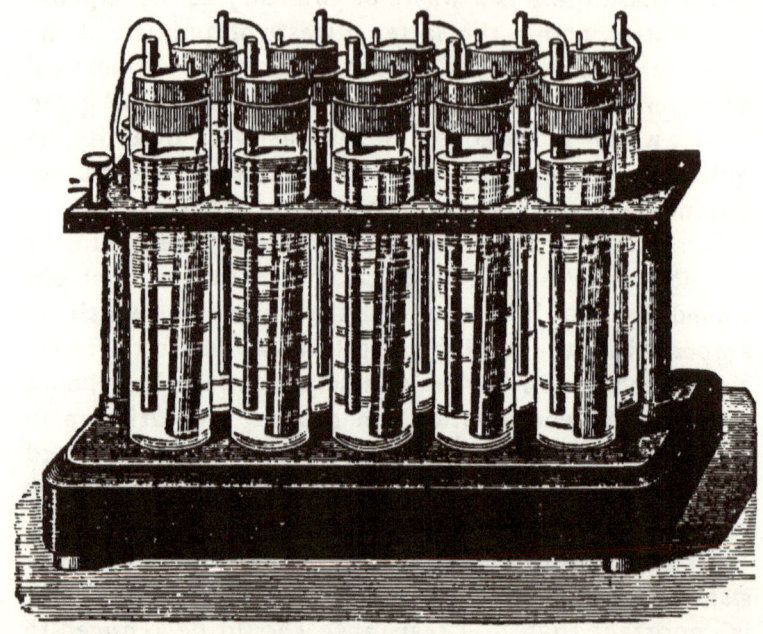

FIG. 16.

close together that the top of the silver electrode can be bent over, and inserted into a hole in the top of the zinc.

Where such methods as the above are not practicable, connection can be made by heavy copper wire or strip, in which the resistance is insignificant. But it is of the utmost importance, in all cases, to have perfect contacts, kept free from oxidation; and this requires frequent, careful inspection.

CHAPTER IV.

MAGNETISM.

The Natural Magnet.—The natural magnet is a hard black stone, which has the property of attracting iron. It derives its name from Magnesia, a country of Asia Minor, where it is supposed to have been first discovered. It was also found at Heraclea, a city of ancient Lydia, and hence called also the heraclean stone. It was known at least five hundred years before the Christian era, being described by Plato and Euripides. It is very rare, but an ore of iron, closely allied to it, known as magnetite, is more abundant, though not always magnetic.

Magnetic Polarity.—No practical use was made of the magnet stone till sixteen centuries after its discovery, when it was found to have the property of assuming a north and south position in the direction of its longer axis, when supported so as to have a free horizontal movement about its centre of gravity. This property was termed *polarity*, from its reference to the earth's poles, and the stone thereafter became known as the *lodestone*—leading stone. It was also observed that iron, rubbed with this stone, acquired its properties of attraction and polarity, and this led to the invention of the mariner's compass.

The Mariner's Compass.—This instrument at first consisted of a thin strip of magnetized iron, named from its shape the *needle*, attached to wood or cork and floated in a vessel of water; a light wooden pointer, attached to it, indicating the ship's course.

In this rude state it became known in Europe early in tne twelfth century, but the exact date and name of the inventor are unknown. A much earlier claim for this invention is made by the Chinese, but does not seem to be well sustained.

The loss of magnetic energy due to the softness of iron seriously impaired the usefulness of the compass, but the subsequent discovery of steel furnished the material for needles much more permanently magnetic. Various improvements followed till it became the perfect instrument which we now have, as represented in Fig. 17. The needle is mounted on a pivot, and at-

Fig. 17.

tached to the under side of a circular card which rotates with it, the margin of which is graduated to thirty-two divisions indicated by pointers, including the four cardinal points N., S., E., W., and also to 360° where great accuracy is required. A circular box with glass cover incloses it, so poised as to maintain a perfect level unaffected by the motion of the ship. The most accurate needles are compound, consisting of several needles connected together. The compass shown in Fig. 17 has eight needles attached to the card, four on each side of the axis of rotation.

The Surveyor's Compass.—This compass differs from the mariner's chiefly in having the needle exposed to view, the graduated circle stationary, and sights and a small telescope mounted above.

The Earth's Magnetic Poles.—The polarity of the needle was found by Gilbert to depend on the magnetism of the earth, which produces north and south magnetic poles by which the needle is attracted; *the polarity of each being opposite to that of the corresponding pole of the needle.* These are not identical with the geographical poles; the north magnetic pole being near the arctic circle, lat. 70° 5′ N., long. 96° 46′ W., and the south near the antarctic circle, about lat. 73° S., long. 154° E., according to Airy's maps, Figs. 18 and 19; the location of the south magnetic pole being only approximate, its position having never been accurately determined.

There are indications of secondary poles also, but their existence and location are not well established; neither is it known whether the magnetic poles are stationary or slowly changing position, as no accurate observations on this point have been made since the discovery of the north magnetic pole by Ross in June 1831; previous to which nothing was known in regard to the location of either magnetic pole.

Declination.—The difference of position between these and the geographical poles produces a deflection of the needle from a true north and south position at all points on the earth's surface except those situated on what is known as the *agonic line*, or meridian of no declination, and this deflection is termed *declination;* the needle's north pole being deflected toward the north magnetic pole, north of the magnetic equator, and its south pole toward the south magnetic pole, south of the magnetic equator. The exact declination at any point is the angle between the vertical plane of the true meridian and that in which the longer horizontal axis of the needle lies at the time of observation.

If the distribution of magnetic force on the earth's surface varied uniformly, it is evident that the agonic line would coincide with the meridian passing through the magnetic and geographical poles, and the declination would vary as the distance east or west of this line; and the magnetic equator, being equally distant from the magnetic poles, would cut the geographical equator at opposite east and west points at an angle of about 20°; and on this equator there could be no declination, horizontal magnetic attraction on each pole of the needle being equal and opposite, while the declination on any geographical meridian not coinciding with the agonic line would vary as the distance north or south of this equator, and attain a maximum of 90° at parallel points adjacent to either magnetic pole. Hence the declination at any point could be calculated from the latitude and longitude if the position of the agonic line were known.

But observation shows that this hypothesis is only approximately true, and that the distribution of magnetic force on the earth's surface is very irregular, as shown by Figs. 18 and 19; and that declination, position of the agonic line, and other magnetic facts can be determined only by actual observation at each point.

MAGNETISM.

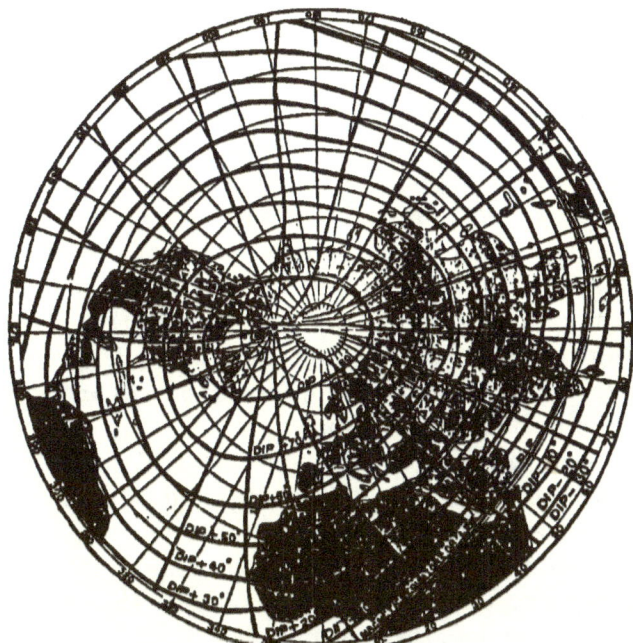

Fig. 18.—Northern Hemisphere.

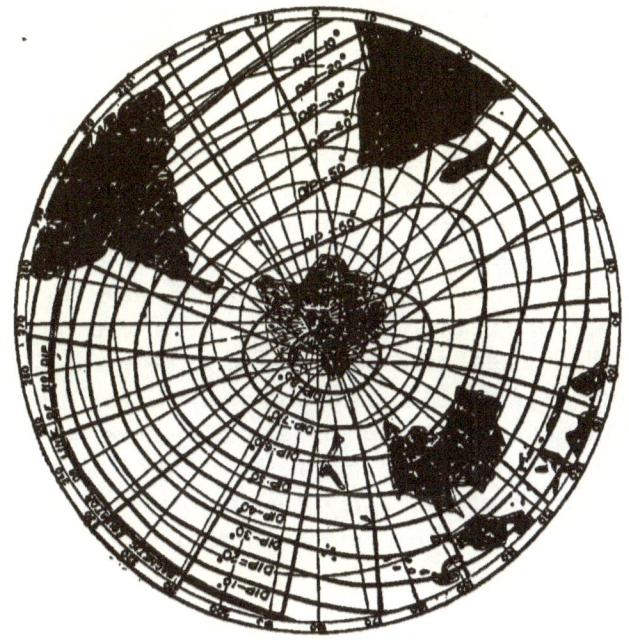

40   DYNAMIC ELECTRICITY AND MAGNETISM.

Inclination or Dip.—At the magnetic equator the position of the needle is parallel with the horizon, vertical magnetic attraction on each of its poles being equal and opposite, but at all points north or south of this line it is inclined at an angle known as its *dip* or *inclination;* its north pole, north of it, inclining towards the north magnetic pole ; and its south pole, south of it, towards the south magnetic pole ; the inclination attaining a maximum of 90° at each, the position of the needle becoming vertical. Hence the compass needle requires a counterpoise sufficient to counteract the dip and keep it in a true horizontal position.

With uniform variation of magnetic force, the inclination at all points between the magnetic equator and poles would vary as the magnetic latitude ; but this is only approximately true, observation being required to determine its value at any point, and also the position of the magnetic equator, and parallels, as well as of the magnetic poles; and as the south magnetic pole has never been definitely located, the vertical position of the needle at that point can only be assumed.

The Dipping Needle.—The inclination is ascertained by an instrument known as the *dipping needle*, constructed with a graduated circle set vertically in the plane of the magnetic meridian, around which a delicately poised needle has a free vertical movement.

Magnetic Maps.—By observation of the declination, dip, and other phenomena at various points on the earth's surface, maps may be prepared which are approximately correct for a limited number of years. The maps of Sir George Airy, Figs. 18 and 19, and of the U. S. Coast and Geodetic Survey, Figs. 20, 21, 22, and 23, have been prepared in this way. The magnetic poles being located with approximate accuracy, the magnetic equator is found by tracing a great circle connecting all points in the equatorial region where the needle maintains a perfect horizontal parallel. This circle, which is very irregular, cuts the geographical equator at opposite east and west points at an angle of about 13°, as shown in Figs. 18 and 19.

The agonic line is found by connecting the points of no declination on a great circle passing through the magnetic poles and cutting the magnetic equator at right angles approximately. Other great circles connecting points of equal declination, and hence called *isogonic lines*, pass also through the magnetic poles and cut the magnetic equator, in like manner, at approximately equal intervals.

Parallels to the magnetic equator, connecting points of equal inclination on the isogonic lines, and hence called *isoclinic lines*, cut the agonic line at approximately equal intervals. All these lines, both of declination and inclination, show great irregularities, the irregularities of the parallels corresponding approximately to those of the magnetic equator.

Terrestrial Magnetism Illustrated.—If a magnetic needle, free to move vertically and horizontally, be brought into the vicinity of a magnetized bar of steel, lying in a north and south position, be moved directly over it from end to end, and also parallel to it at a short distance on each side, it exhibits all the phases of declination and dip found on the various parts of the earth's surface, as already described, but in a more regular manner; which is strong proof that the earth is a great magnet, as already stated, with curving lines of force radiating in all directions from its magnetic poles, like other magnets, as described hereafter; thus accounting in a most satisfactory manner for the phenomena of declination and dip.

These are the lines represented in part on the maps,

the needle being merely the instrument by which they are traced.

Magnetic Intensity.—The intensity of this magnetic force constantly increases from the magnetic equator to each magnetic pole, and is represented at any point by the forces producing the declination and dip; the former representing the horizontal component of the intensity, and the latter the vertical; the total intensity being ascertained by dividing the horizontal force by the cosine of the angle of dip. Hence, representing the total intensity by F, the horizontal force by H, and the angle of dip by θ, we have the usual standard formula $F = \dfrac{H}{\cos \theta}$.

There are two methods of ascertaining the relative values of the horizontal force at different points, known respectively as the methods by *oscillation* and by *deflection*.

Magnetic Force Ascertained by Oscillation.—The oscillations of the needle, when forcibly deflected from its position of rest, are accomplished, like those of the pendulum under the influence of gravity, in nearly equal times, though constantly decreasing in amplitude; and the square of the number of oscillations accomplished in a given time, which in the pendulum is proportional to the force of gravity, is, in the needle, proportional to the horizontal magnetic force. Hence if a represent the number accomplished in a given time at any point on the earth's surface, and b the number accomplished in the same time by the same needle at any other point, the relative values of this force at the two points are as a^2 to b^2.

Magnetic Force Ascertained by Deflection.—The horizontal force by which the needle is brought to rest in the plane of the magnetic meridian is the resultant of two forces, one tending to rotate it into an east and west position, represented by the sine of the angle of declination, and the other into a north and south position, represented by the cosine, while the resultant force is represented by the hypothenuse of a right-angled triangle, of which the sine and cosine form the remaining sides (see Fig. 45, page 121); the position of the needle coinciding with that of the hypothenuse, in which the forces are in equilibrium. The relative value of the east and west force, by which the needle is deflected, is to that of the north and south force, as the ratio of the sine to the cosine, represented by the tangent. Hence the total horizontal force of the earth's magnetism, at any point, multiplied by the tangent of the angle of declination gives the deflective force at that point.

Absolute Magnetic Intensity.—The relative magnetic intensity being derived, as shown, from division of the horizontal force by the cosine of the angle of dip, if the absolute value of this force, in C. G. S. units, at any point is ascertained, the absolute intensity can also be ascertained. To accomplish this two observations are necessary with a needle whose magnetic *moment* or force by which it resists deflection is known; a quantity ascertained by multiplying the strength of either pole by the length of the needle (or magnet). One of these observations, made by oscillation, determines the *product* of this moment by the horizontal force; and the other, made by the special deflection of a small needle by the same needle (or magnet) used in the first observation, determines the *quotient* of the moment by the horizontal force; and dividing the product by the quotient and taking the square root of the result gives the absolute horizontal force.

Previous to 1830 observations on magnetic intensity were made by oscillation of the dipping needle, but this method was found to be inaccurate and the observations unreliable. The discovery of the method of expressing this intensity in absolute measure was first made by Gauss in 1833, and the portable magnetometer (described in the latter part of this chapter), an important aid in such measurement, was constructed by Weber in 1836.

The number of oscillations made in a given time by different needles, or magnets, varies as the length, weight, form, and polar strength of each, and as the strength of the magnetic field in which it is placed. Hence, in comparing observations made by different instruments, it is necessary to correct any errors which may arise from such variation. It is also important to prevent errors due to loss of magnetism, by frequent testing, and remagnetizing when necessary.

Parallels to the magnetic equator, connecting points of equal magnetic intensity, and hence called *isodynamic lines*, are traced on maps representing either the horizontal or the total intensity, as shown in Fig. 20.

Biot's Law.—The magnetic intensity at any point on the earth's surface varies with the magnetic latitude; to which it is approximately proportional. The magnetic force, emanating from the magnetic poles and radiating in curves as already stated, not only on the surface but into the surrounding space in all directions, varies inversely as the square of the distance from either pole, except as modified in the manner already shown. Hence the intensity is greatest at the magnetic poles and least at the magnetic equator, and may be ascertained approximately at any point by Biot's law, which, representing the magnetic latitude by l, makes the intensity proportional to $\sqrt{1 + 3 \sin^2 l}$.

Origin of Terrestrial Magnetism.—The origin of terrestrial magnetism and its peculiar phenomena is to be found in the reciprocal relations of magnetism and electricity as explained in the next chapter, each being capable of producing the other.

Electric terrestrial phenomena have been described in the author's " Elements of Static Electricity," Chap-

ISODYNAMIC MAP OF THE UNITED STATES FOR 1885.

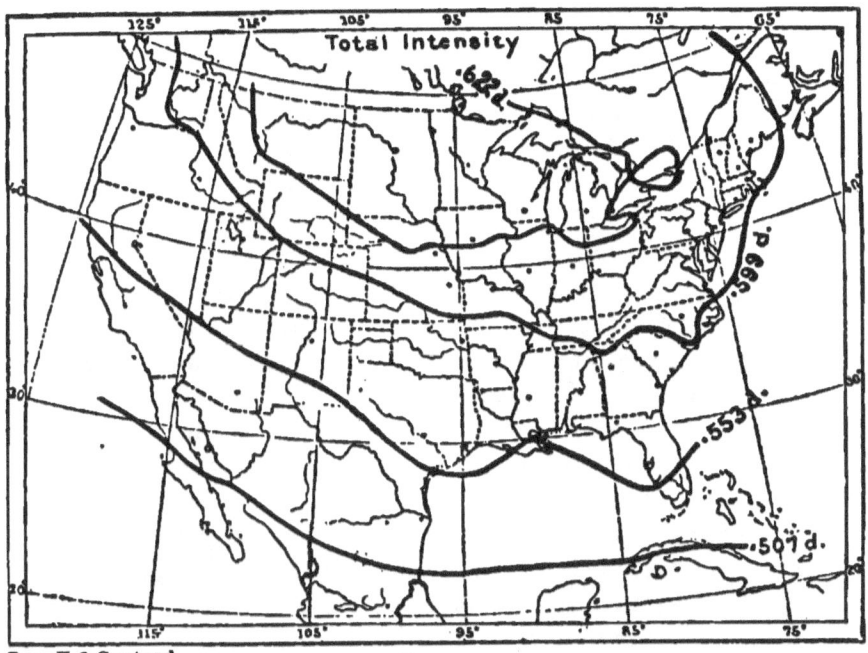

From U. S. Coast and Geodetic Survey.   FIG. 20.

ters XII and XIII, where it has been shown that difference of electric potential between different parts of the earth's surface and atmosphere is apparently the result of difference of temperature, modified by the unequal distribution of land and water; hence the magnetic terrestrial phenomena which we have been considering may be regarded as the result of the electric phenomena; and the peculiar phases of each, as indicated by geographical position and otherwise, leave no doubt of their intimate relationship; so that whether the magnetic phenomena be regarded as a result of the electric, or the reverse, both are undoubtedly dependent on the same physical influences.

Secular Variation.—Observation shows that the special phases of terrestrial magnetism are subject to great variation in respect to time as well as geographical position; such variation being of three kinds, secular, annual, and diurnal. The first embraces long terms of years known as *secular periods*, whose length is determined by the time in which a complete cycle of changes occurs. The discovery of this variation is due to Gellibrand, an English electrician, and was first published in 1635.

The agonic line and the isogonic lines are constantly changing position, slowly vibrating between widely separated eastern and western limits; hence the declination at any point shows a corresponding variation between eastern and western maxima; and the time occupied by the agonic line or any isogonic line in passing from its eastern or western limit, on any magnetic parallel, until its return to the same limit again, or by the magnetic needle, at any point, in vibrating from its eastern or western maximum declination, or elongation, until its return to the same declination again, constitutes a secular period.

When the declination has attained a maximum, it becomes apparently stationary, change in the opposite direction being for some years imperceptible, after which the mean annual variation steadily increases for a term of years till the declination becomes zero; a corresponding decrease of annual variation then occurs till it again becomes imperceptible, and the declination apparently stationary, at the opposite maximum; there is then a return through a similar series of variations to the original maximum.

This variation in rate of declination during a secular period has its exact analogy in the similar variation of rate found in a vibration of the pendulum.

The length of a secular period is not definitely known, as sufficient time has not yet elapsed since observations were first made at any point for a complete cycle of changes to occur. It varies considerably in different parts of the earth; for the United States it is estimated at from 250 to 350 years, and for Paris at about 470 years; the earliest observations having been made there, dating back to 1540.

Secular vibration does not necessarily imply a change in the direction of the needle from west of north to east, or the opposite, which occurs only within the range of the agonic line; at all points east of that range the needle always points west of north, and at all points west of it east of north. Neither is it to be understood that the vibration of the needle within this range differs from that outside of it; the agonic line is simply the boundary between east and west declination, and at all points within its range the needle changes direction, once from east to west of north, and once from west to east, during the secular period, according as this line vibrates past each point in either direction.

This would imply that the general vibratory movement in either direction is simultaneous at all points, increase of east declination and decrease of west declination, or the opposite, occurring everywhere at the same time; and that when either has attained its maximum elongation the other has attained its minimum, all the lines having swayed to the west or to the east simultaneously. But this is never strictly true except within a very limited area; declination may have at-

48   DYNAMIC ELECTRICITY AND MAGNETISM.

tained its maximum or minimum at a remote point east or west of any isogonic line, and the opposite phase set in long before the same change occurs at intermediate points; so that it may be increasing or diminishing in opposite directions at the same time on the same parallel.

Secular Variation in the United States.—It is found that this magnetic wave has thus swept across the American

ISOGONIC MAP OF THE UNITED STATES FOR 1885.

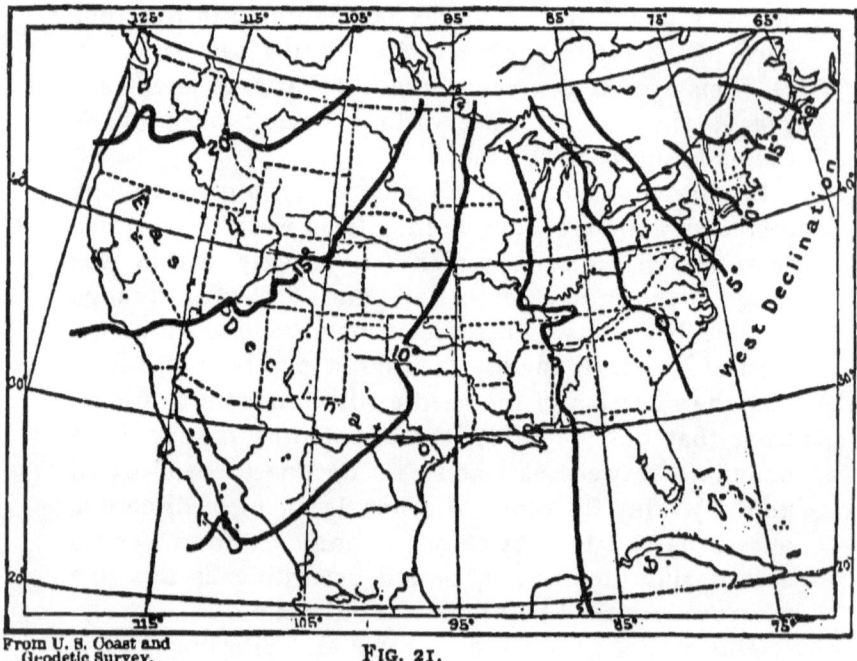

From U. S. Coast and Geodetic Survey.   FIG. 21.

continent from east to west since observation began, and that its return eastward is now setting in. East elongation had attained its stationary phase, followed by reversal, at Halifax, N. S., in 1713, at Eastport, Me., in 1749, at Boston in 1780, at New York in 1799, at Pittsburg in 1808, at Cincinnati in 1815, at Chicago in 1832,

MAGNETISM. 49

at Salt Lake City in 1873, and will attain it at San Francisco, as computed, in 1893.

In 1890 declination, throughout the interior of the United States, was tending westward; west declination increasing and east declination diminishing, while at the extreme eastern and western points it had become stationary and the opposite phase was setting in; west declination beginning to decrease on the east coast, and east declination to increase on the west coast.

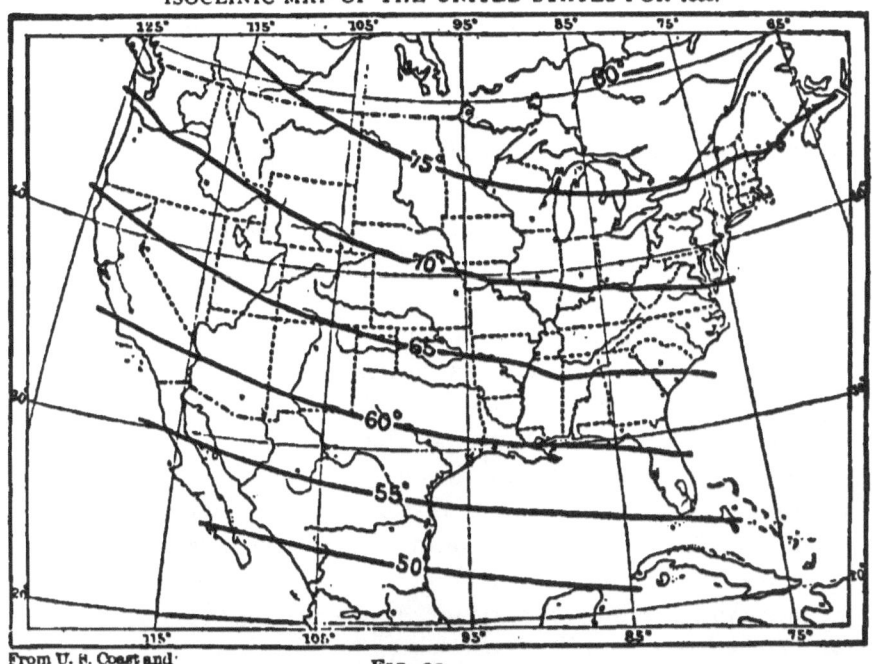

ISOCLINIC MAP OF THE UNITED STATES FOR 1885.

From U. S. Coast and Geodetic Survey.   FIG. 22.

The secular periods of dip and magnetic intensity are apparently the same as those of declination.

The earliest reliable observations for the United States are those given in Halley's chart for the year 1700. The agonic line, moving eastward, then passed a little east of Charleston, S. C.; declination at that point Jan. 1st. being 36′ E. In 1790 declination had attained its maximum eastern elongation, $4° 54\tfrac{6}{10}′$ E., followed by decrease till about Jan. 1, 1890, when the agonic line was a little west of Charleston, declination being $4\tfrac{2}{10}′$ W.

Annual and Diurnal Variations.—The annual and diurnal variations are apparently due to change of temperature. The annual maximum variation occurs in summer and the minimum in winter, corresponding respectively to the months of greatest and least change of temperature; the diurnal maximum during the day and the minimum during the night, corresponding respectively to the hours of greatest and least change of temperature. These variations are very slight, that of diurnal declination being greatest, ranging from 6′ to $10\tfrac{1}{2}′$, as observed at Philadelphia and at London; while the annual does not exceed $1\tfrac{1}{2}′$.

The electric terrestrial phenomena already referred to show annual and diurnal variations corresponding in time and amount to the magnetic.

The Eleven Year Period.—It is found that the greatest magnetic diurnal variations take place at regularly recurring periods of about eleven years each, corresponding to the periods of greatest solar disturbance, as indicated by the sun-spots, and of most frequent occurrence of the electric phenomena known as the aurora; minimum periods recurring at intervening epochs of eleven years.

Magnetic Storms.—Unusual perturbations in terrestrial magnetism often occur, known as *magnetic* or *electric storms*, lasting usaully only a few hours, though sometimes much longer. They are indicated by sudden and unexpected deflections of the needle, and great and rapid fluctuations from its normal position, and also by other magnetic and electric disturbances, which occur simultaneously over extended areas, often embracing distant parts of the globe. They usually occur in connection with the aurora, and are accompanied by electric currents in the earth, each phenomenon being doubtless due to the same cause, as explained in "Elements of Static Electricity," Chapter XV.

Cosmic Variation.—There is also slight magnetic variation due to solar and lunar influence, which may properly be termed *cosmic*. That due to solar influence depends on the rotation of the sun on its axis, and hence has a corresponding period of about 26 days. That due to lunar iufluence exhibits two maxima and two minima during each lunar month, the difference between which at Philadelphia is about 27", and at Toronto about 38".

Exact Observation.—In view of these numerous variations it is evident that the magnetic needle, when light and delicately poised, so as to be sensitive to the slightest change, is in a state of constant tremulous motion, and never absolutely at rest ; so that the record of an observation, to be of true scientific or even practical value, must specify the exact time, limited to the day and hour when made, as well as the exact location. This is especially true of declination, which is of the highest practical importance in surveying and navigation, often involving important legal controversies.

Secular Variation at Washington.—Observation at Washington began about 1790, at which date the agonic line passed through it. In 1797 this line had attained its eastern limit, and declination at Washington its eastern maximum, being 30' E. Jan. 1st. The agonic line has since been moving westward, and passed through Washington again in 1803.

Fig 23 shows the position of this line in the United

52 DYNAMIC ELECTRICITY AND MAGNETISM.

MAP SHOWING THE POSITION OF THE AGONIC LINE IN THE UNITED STATES AT FOUR DIFFERENT EPOCHS.

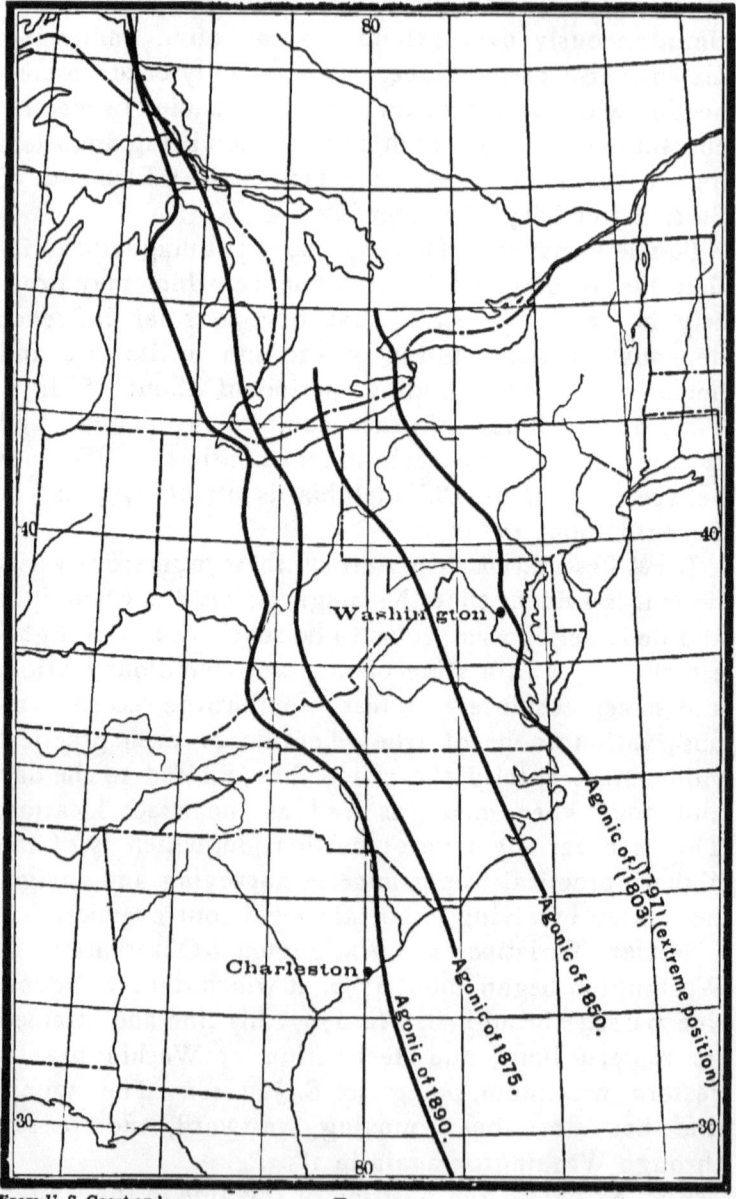

From U. S. Coast and Geodetic Survey.

FIG. 23.

MAGNETISM. 53

States at four different epochs, including that of its eastern limit. In 1810 the declination was 12' W.; in 1830, 39' W.; in 1850, 1° 58$\frac{8}{10}$' W.; in 1870, 2° 55$\frac{8}{10}$' W.; and in 1890, 4° 15$\frac{8}{10}$' W.

The mean annual variation in declination from 1790 to 1810 was 0.6'; from 1810 to 1830, 1.35'; from 1830 to 1850, 3.99'; from 1850 to 1870, 2.85'; and from 1870 to 1890, 3.99'; the average for the entire period from 1790 to 1890 being 2.55$\frac{3}{8}$'.

Observations on the dip and magnetic intensity in the United States date back to the latter part of the last century; but, on account of imperfections in the instruments and methods of observation in general use for this purpose previous to 1838, observations made before that date are not considered very reliable. The dip at Washington was then 71° 13'; for the next 22 years there was alternate increase and decrease, the maximum being attained in 1845, when it was 71° 34'. In 1860 it was 71° 20'; it then steadily declined, with slight alternation, at a mean annual rate of about 1.75', and in 1890 was 70° 24'.

The total magnetic intensity at Washington in 1840 was 0.61923 of a dyne. It decreased from that date to 1850, when it was 0.61370; increased to 0.61877 in 1865, and decreased to 0.60863 in 1885.

Secular Variation at San Francisco.—The declination at San Francisco in 1790 was 13° 6' E.; in 1810, 14° 6' E.; in 1830, 15° E.; in 1850, 15° 47$\frac{4}{10}$' E.; in 1870, 16° 20$\frac{4}{10}$' E.; and in 1890, 16° 34$\frac{8}{10}$' E.

The mean annual variation in declination from 1790 to 1810 was 3'; from 1810 to 1830, 2.7'; from 1830 to 1850, 2.37'; from 1850 to 1870, 1.65'; and from 1870 to 1890, 0.72'; the average for the entire period from 1790 to 1890 being 2.088'.

There are no accurate data from which variation in dip and total magnetic intensity on the west coast of the United States, for an extended period, can be ascertained, reliable observation on these phenomena being very recent. The dip at San Francisco in 1885 was $62° 15'$, was thought to have just passed its maximum and to be slowly decreasing; and the total magnetic intensity for the same date was about 0.5456.

Artificial Magnets.—Various metals besides iron and steel can acquire magnetism, but only in a slight degree, especially nickel and cobalt, also chromium, cerium, and manganese; and Faraday found that all substances apparently are susceptible of magnetic influence, as might be inferred from the magnetism of the earth itself. But steel of high temper is the only metal capable of both acquiring and retaining magnetism to a sufficient degree for practical purposes, hence all permanent artificial magnets are made of it; and magnetized steel, like the natural magnet, is capable of magnetizing steel or iron brought into contact with it without impairing its own magnetism.

Steel magnets may be of any convenient size and form, but are usually made either straight or U-shaped, 2 to 12 inches in length, $\frac{1}{2}$ inch to an inch or more in width, and $\frac{3}{16}$ to $\frac{3}{8}$ of an inch in thickness. They may be magnetized to a certain degree by simple contact with a natural or artificial magnet, also by electricity or, more effectually, by the electro-magnet, as explained hereafter; or, if straight, by placing them for a considerable time parallel to the line of inclination in the plane of the magnetic meridian.

This process may be hastened by concussion or, in the case of a wire, by torsion, both indicating that magnetism is a molecular effect, the molecules being under a strain in this position to which they yield more rapidly when assisted by the concussion or torsion.

A common method is represented by Fig. 24. Two magnets having their opposite poles in contact and

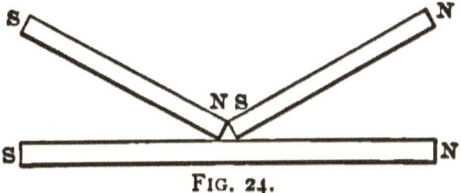

FIG. 24.

resting on the bar to be magnetized, at its centre, are drawn apart to its opposite ends and brought together again at the centre repeatedly an equal number of times on its opposite surfaces till it becomes magnetically *saturated*, the final movement on each surface terminating at the centre. The polarity acquired by the bar at each end is opposite to that of the magnetic pole brought into contact with it. Bars of the U or horseshoe form are magnetized in a similar manner by taking the centre of the bend as the point for beginning and terminating.

Another method is to interpose a non-magnetic body between the magnetizing poles and move both alternately from end to end over each surface an equal number of times.

Magnetic Saturation.—By magnetic *saturation* is meant the full quantity of magnetism which the bar is capable of retaining. It may be magnetized above the point of saturation, when it is said to be *super-saturated*, but the extra magnetism thus acquired is rapidly dissipated.

The Armature.—The magnetism of straight bar magnets and magnetic needles becomes impaired in the course of years, and the needles especially require to be remagnetized to maintain the requisite strength. The U and horseshoe magnets are each furnished with a piece of soft iron connecting the poles, and held there by magnetic attraction. It is known as the *armature* or *keeper*, from its supposed ability to prevent magnetic loss by completing the magnetic circuit. The chief advantage of the U or horseshoe form is in the concentration of the magnetic attraction of both poles on the same armature, which is itself thus temporarily magnetized and has poles of opposite magnetism to those by which it is attracted, which increases the force of attraction.

Two bar magnets placed near each other, side by side, can also have armatures with advantages similar to those just mentioned. The term armature is also similarly applied, in the construction of electro-magnetic apparatus, to a piece of iron attracted by a single pole; also in a special, technical sense, in the construction of the dynamo, as explained hereafter.

Laminated Magnets.—Magnets of either form mentioned are often made of a number of thin bars, separately magnetized and bound together with their similar poles in contact. They are known as *laminated*, and have greater magnetic strength than those having the same amount of steel in a single piece, an effect probably due to the partial suppression of magnetic eddy currents, to which the steel massed in a single thick piece is liable. Such currents occurring within the mass, like electric "local action" within the battery cell, tend to neutralize and reduce magnetic potential difference, while their external effect is lost. But the laminated structure tends to confine the lines of force to the laminæ and give them a normal direction toward the poles. These eddy currents were first observed by Foucault, and hence are termed *Foucault currents*. In large electro-magnets special construction is required to guard against their deleterious effects.

Magnetic Loss.—In addition to the loss by gradual

dissipation, already referred to, steel magnets lose their magnetism by sudden or violent perturbations, such as concussion, a white heat, or such extreme cold as $-100°$ C., which indicates, as before, that magnetic change is dependent on molecular change.

Magnetized iron rapidly loses its magnetic energy, especially iron of the softer grades, but retains a small amount known as *residual magnetism ;* a similar amount being often acquired in the process of the manufacture of iron machinery.

Portative Force.—The attractive property of the magnet is manifested in sustaining pieces of iron or steel suspended from it, both by direct contact and through the medium of other pieces, so that a number of small pieces, as nails or needles, may be suspended from each other. This property is known as its *portative force.* If connection with the magnet be severed, the iron quickly loses its magnetism, but the steel retains it to an extent governed by its mass and temper.

Magnets of the U or horseshoe form can sustain weights attached to their armatures to the amount, in some cases, of twenty times their own weight, and a little lodestone mounted in Sir Isaac Newton's signet-ring could sustain two hundred times its own weight.

The portative force does not vary in the direct ratio of the mass; the proportion of force to mass being greater in small than in large magnets. The following rule is given by Bernoulli:

Let p represent the force, w the weight of the magnet, and a the quality of the steel and method of magnetizing; then $p = a\sqrt[3]{w}$.

The weight sustained varies also somewhat as the area of surface contact between the magnet and its armature, and the portative force is gradually increased by frequent additions to the load, but this increase is lost by sudden separation of the armature from the magnet.

Polar Attraction and Repulsion.—If the similar poles of two magnetic needles or magnets, free to move, are placed in mutual proximity they repel each other, but if their dissimilar poles are so placed they attract each other; from which is derived the law that *like magnetic poles repel and unlike attract each other.*

From this it will be seen that the poles of the needle are opposite in polarity to those of the earth by which they are attracted, but for convenience the pole which points north is termed the north pole, and that which points south the south pole; though they are sometimes more appropriately termed the *north-seeking* and *south-seeking* poles. Their initial letters, N and S, are used to distinguish them, magnets being usually marked on the north pole only, by N or a cross-line; hence the expression "marked pole" is sometimes used to distinguish it from the south or "unmarked pole."

It is impossible to produce one kind of polarity without at the same time producing its opposite. This becomes manifest in a very striking manner when a magnet is broken, the pieces assuming opposite polarity on opposite sides of the fracture and at opposite ends, each becoming a perfect magnet. If the parts be pressed closely together in their original position these poles disappear, leaving only the poles at the original ends; and the same thing occurs if the opposite poles of two rectangular magnets of the same cross-section be pressed together so as to make perfect joint.

A magnetic pole may have sufficient strength to overcome by induction the repulsion of a weaker one of similar polarity and produce attraction when brought sufficiently close, while at a greater distance where the induction is less there is repulsion.

MAGNETISM. 59

Unmagnetized iron or steel is attracted by either pole of the magnet, and apparently attracts either pole, but only as the result of reaction, being itself magnetically passive.

Magnetic Lines of Force.—If a sheet of paper be placed over a bar magnet and iron filings dusted over it, the sheet being lightly tapped, the filings arrange themselves in curves corresponding to the lines of force emanating from the poles, as shown in Fig. 25. Each filing becomes itself a magnet by induction, and their

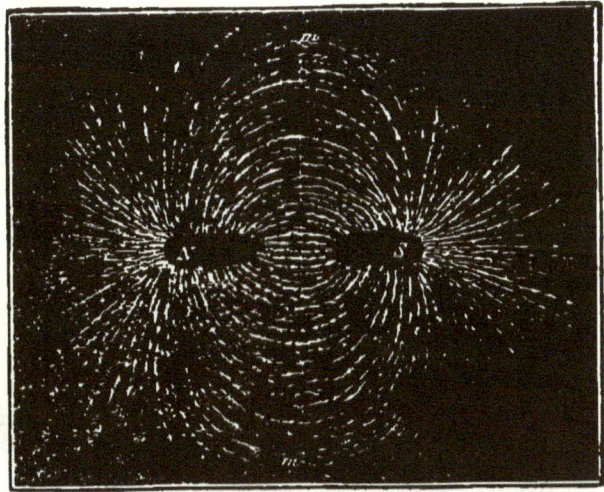

FIG. 25.

dissimilar poles being, mutually attracted, attach themselves to each other, while their similar poles are mutually repelled; hence poles of the same name all point in the same direction.

The lines of force, radiating from each pole of the magnet, being mutually repelled and attracted toward the opposite pole, are under the influence of two forces, each varying inversely as the other, the one urging them directly forward and decreasing as the square of the distance from each pole increases, the other attracting them toward the opposite pole, and increasing in the same ratio as the first decreases, becoming greater as the distance to the opposite pole lessens and the distance from the originating pole increases; hence the resulting curves are formed as indicated by the filings.

The space inclosed by lines of force is termed a *tube of force*.

Magnetic Field.—The filings show the lines of force only in the plane of the sheet of paper, which may be placed at any angle at which they can be sustained in position; while close to the poles they stand on end nearly at right angles to the paper. Hence it becomes evident that the lines of force inclose the magnet in the form of a spheroid which is cut by the plane of the paper and shown in longitudinal section. The friction of the filings on the surface, and their weight, inertia, and tendency to mass together, prevent free movement,

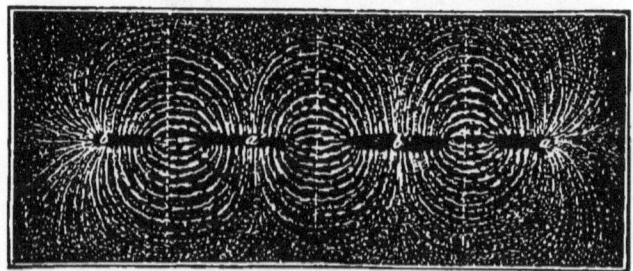

FIG. 26.

so that they indicate the actual position of the lines of force very roughly. Those lines must be understood to fill the entire space inclosed by the spheroid, constituting what is known as the *magnetic field*; a more accurate conception of which would be such a figure as might be supposed to form itself around a magnet suspended in an atmosphere of iron vapor.

In Fig. 26 the filings represent what are known as

MAGNETISM. 61

consequent poles, which may result from imperfections in the temper of the steel or in the method of magnetizing. Such poles may also be produced in a thin bar of highly tempered steel by touching it at several points with a magnet. The result in either case is similar to that of several short magnets joined together by their opposite poles.

In Fig. 27, A and B, the filings show repulsion and

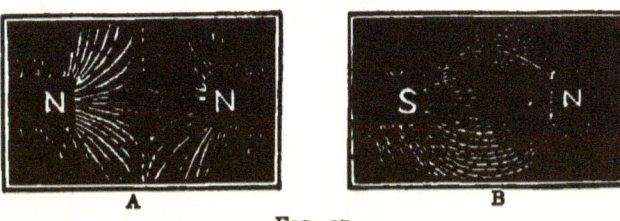

FIG. 27.

attraction in a very instructive manner. A represents like poles in proximity with the lines curving away from each other, while B represents unlike poles in a similar position with the lines curving towards each other.

In Fig. 28 is shown at A how the opposing lines of force from like poles produce mutual repulsion as already described, while at B the curving lines from unlike poles interlock and produce mutual attraction.

Since these lines do not radiate into vacant space, but into the air, it may safely be assumed that the air is the medium by which the magnetic field is formed, and that it becomes magnetized in a manner similar to that of the iron filings by which the field is represented in section; which accounts rationally for the observed attraction and repulsion. This hypothesis is strengthened by the consideration that it is impossible for energy, in any form, to exist independent of matter, so that the field could not be formed in an absolute vacuum; magnetism itself being doubtless an effect of energy acting on matter. Hence we must either assume that the matter, in this case, is the air of which we have actual

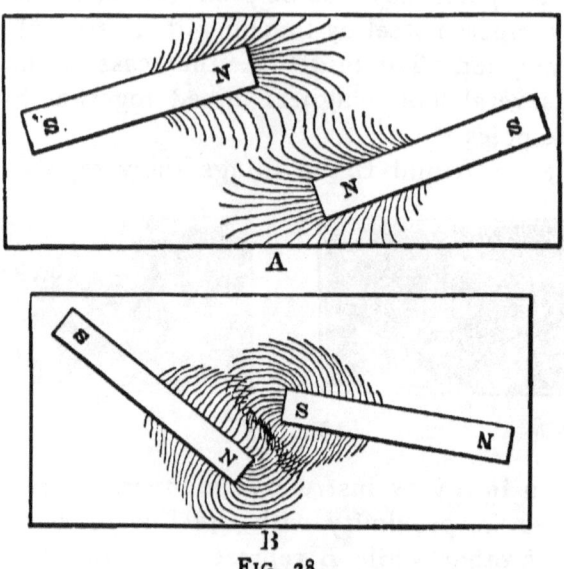

Fig. 28.

knowledge, or the hypothetical ether whose actual existence has never been demonstrated.

The interposition of any substance in the magnetic field except that of magnetic bodies, as iron or steel, or those termed diamagnetic, as bismuth and copper, offers no obstruction to magnetic induction; attraction or repulsion taking place through all other bodies with the same facility as if they were not present: while steel or iron, especially soft iron, absorbs and diverts the lines of force through its own substance in proportion to its mass, extent, and relative position. Iron may therefore be used as a shield against magnetic influence.

Diamagnetic bodies also offer slight obstruction, but

in an opposite sense to that of iron or steel, resisting or turning aside lines of force instead of absorbing them.

Form of Magnets.—The forms of the magnet already described are the most convenient for practical use; but iron or steel in any form may be magnetized, becoming polar normally in the direction of its longer axis, which would be true of magnets having the form of the spheroid, cylinder, or ellipse, as well as of the more common forms; but in such forms as the sphere or circle, having equal radii, the separate poles are not distinguishable, but must be supposed to exist and neutralize each other.

Opposite surfaces of sheet iron or steel may be so magnetized as to acquire opposite polarity; such magnetic distribution being known as *lamellar*, in distinction from the ordinary distribution, which is termed *solenoidal*, and such sheet magnets are known as *magnetic shells*.

Magnetic Penetration.—The depth to which magnetism penetrates depends somewhat on the degree of magnetization and the size of the bar in cross-section. It is strongest in the outer layers, as may be shown by removing them with sulphuric acid, when the magnetism will be found to become constantly weaker as the central core is approached. The same may also be shown by magnetizing bundles of thin steel plates bound together and gradually removing the outer ones; also by means of a magnetized steel tube, whose magnetism is found to be nearly equal to that of a solid bar of the same cross-section.

Location of the Poles.—The poles have been thus far assumed to be at the ends of the magnet's longer axis, but this is practically true only of long, thin, narrow magnets uniformly magnetized throughout; their true location is a little inside of the ends, the distance varying inversely as the mass in cross-section, so that in thick magnets it becomes noticeable.

Paramagnetic and Diamagnetic Bodies.—Faraday proposed to call such bodies as are capable of assuming the magnetic properties of attraction and repulsion, as iron, steel, nickel, and cobalt, *paramagnetic*, while bodies which are repelled by either magnetic pole, as bismuth, antimony, copper, and phosphorus, he proposed to call *diamagnetic*. But these terms have been adopted only in part, the utility of "paramagnetic" especially being questioned.

Some writers accept the term diamagnetic as applied to the latter class of bodies, and designate the former as magnetic, a usage which is simpler and more convenient.

Magneto-Crystallic Induction.—Bodies, whether transparent or opaque, having a crystalline structure are influenced by magnetic action differently from bodies which do not possess such structure, and diamagnetic crystalline bodies differently from magnetic crystalline bodies. Such a body when suspended subject to magnetic induction is most strongly influenced in a certain direction known as its magne-crystallic axis, which, in crystals having cleavage, is usually at right angles to the plane of cleavage. This axis, according to Tyndall, seems to lie in the direction of the crystal's greatest density, and magnetic crystals, free to move, take position with this axis in the direction of the lines of force, while in the position assumed by diamagnetic crystals it is at right angles to those lines.

The whole subject is imperfectly understood, and opinions in regard to it are conflicting; and in its present aspect it must be regarded as of secondary importance.

Magnetism as a Mode of Molecular Motion.—The magnetic phenomena thus far observed indicate that magnetism is closely related to the molecular constitution of the magnetized body. The effects of extreme heat or extreme cold, of concussion and torsion in producing or destroying magnetism, all of which affect the molecular constitution, have already been noticed; it is also found that when iron filings are closely packed in a tube and magnetized the mass exhibits all the magnetic properties of a bar magnet, but if disturbed by being shaken up the magnetism disappears. The filings in this case may roughly represent the molecules of a solid bar, and the magnetic loss is analogous to that produced in such a bar by concussion. If the tube is of glass the filings can be seen to arrange themselves lengthwise, with similar poles all turned in the same direction, as already observed in the curved lines of filings, constituting magnetic series, from which results the polarity of the mass, as in magnets with consequent poles.

It is found that magnetizing a steel bar produces a slight change in its form, the bar becoming a little longer, and its other dimensions being correspondingly reduced; an effect attributed to a change of position in its molecules. In the unmagnetized bar the molecules may be supposed to be massed together without order, but under the magnetic influence to arrange themselves symmetrically in the direction of their longer axes, with similar poles in the same direction, like the filings, the results being the change of form mentioned above and the polarity of the bar. This theory receives further support from the slowness with which steel acquires magnetism, as compared with soft iron, and its power of retention, while the iron both acquires and loses it rapidly; this quality in the steel being attributable to its rigidity, which resists change of position in the molecules, while those of soft iron easily yield to such change Further proof of the same character is found in the fact that when a bar is magnetized suddenly by electric action a clink may be heard in it, both at the beginning and end of the process, which can be satisfactorily accounted for only on the theory of molecular action.

The above phenomena clearly indicate molecular change of position as a result of magnetization, and hence motion; but if the motion should cease when the molecules have assumed symmetrical position, magnetic action should also cease, for it would be absurd to suppose that this action could continue when the motion which gave rise to it ceased, and that it should be the result of mere symmetrical arrangement, the molecules thereafter remaining quiescent: but we find, on the contrary, that it is then at its maximum, and, in steel, remains permanent for years. Hence we must infer a corresponding maximum and continuity of molecular motion.

The character of this motion cannot be known. We may suppose each molecule to oscillate in the direction of either its longer or shorter axis, or to rotate around either axis, or to have a vortical motion, or to combine two or more such motions; but it is a reasonable inference that this motion, whatever its character, is similar and uniform in each molecule, so that there is no interference between them such as would result from a difference either in their motions or position. And this condition may be supposed to constitute the difference between the magnetized and unmagnetized bar; or more explicitly, that this motion is itself that which we term magnetism.

Hence if we could be endowed with some superior sense, capable of penetrating the magnet and revealing its separate molecules and their motions, we should probably see, instead of a quiescent body, a quivering mass of innumerable atoms, each moving with inconceivable rapidity, and all in a uniform manner and in obedience to a common impulse.

This theory of magnetism has the sanction of Clerk Maxwell, Hughes, and others, and is in accordance with the similar theories in regard to heat and electricity among whose advocates Tyndall is prominent. And as we have seen that different kinds of motion may be combined in the same molecule, without interference, as in larger masses, we may attribute the heat to one kind, the electricity to another, and the magnetism to a third, all being different manifestations of that universal energy which is inherent in all matter.

The theory of magnetism being a fluid, or two dissimilar fluids, once so popular, has now become obsolete, and cannot be sustained on rational grounds in the light of recent investigation.

Analogy between Magnetic and Electric Phenomena.—The close analogy between many of the phenomena of electricity and magnetism indicate that both are closely allied, if not identical, as will appear more fully in the next chapter. But it may here be noticed that opposite magnetic polarity has its analogy in opposite electric polarity; that in both cases the opposite kinds are coexistent and neutralize each other; that magnetic distribution is influenced by the form of bodies in a manner similar to that of electro-static distribution; and that magnetic attraction and repulsion has its analogy in electric attraction and repulsion. But in all these analogies there is a well-defined difference observable which easily distinguishes the magnetic from the electric phenomena.

Coulomb's Torsion Balance.—A full description of this instrument and its application to the measurement of electric force may be found in the writer's "Elements of Static Electricity." It is used in a similar manner to measure magnetic force, as shown in Fig. 29. It consists of a circular glass case, with a vertical cylinder projecting from the cover, having at its upper end a graduated circle, with a pointer to move round it attached to a milled head, from which is suspended, by a fine wire, a magnetic needle, with its poles opposite a circle on the case, graduated to correspond to the one above.

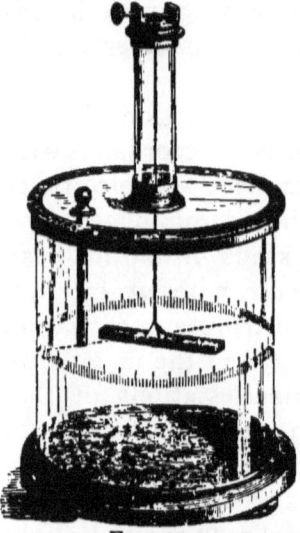

FIG. 29.

The following experiment made by Coulomb shows its use: Zero of the lower scale being brought opposite one of the needle's poles, and accurately adjusted in this position by comparison with a copper needle of equal weight suspended by a thread, and the upper scale being adjusted with its zero opposite the pointer, it was ascertained, by a preliminary trial, that the milled head required to be turned 35° to produce 1° of deflection in the needle; hence magnetic force, in this case, was to torsion as 1 to 35. A magnet was then inserted, as shown, with its pole close to the similar pole of the needle, and was found to produce 24° deflection; hence its force, as ascertained above, was 35 times 24°, to which must be added the 24° of torsion, giving $24° \times 35 + 24° = 864°$ as the "*torsion equivalent*" of magnetic repulsion with the poles 24° apart.

The milled head was then turned so as to reduce this distance one half (12), requiring 8 complete turns, which gives $8 \times 360° = 2880°$. But the 12° remaining at bottom represented a force equal to half the original "torsion equivalent," or 432°, which must be added in, giving 2880° + 432° = 3312°, nearly *four times* 864°, as the "torsion equivalent" of magnetic repulsion at *one half* the distance.

In like manner it can be shown that any reduction of distance, algebraically represented by a, would require a "torsion equivalent" equal to a^2; hence magnetic force is thus proved to *vary inversely as the square of the distance*.

The inaccuracy observable in the arithmetical result is accounted for by inaccuracies in the instrument, and in the angular measurement adopted, which are fully explained in "Elements of Static Electricity" referred to above.

The Gauss-Weber Portable Magnetometer.—This instrument, used for measuring the horizontal force of the earth's magnetism at any point, as shown by the magnetic declination, is constructed as follows: A bar magnet of convenient size is suspended horizontally, from a vertical standard, by an unspun silk fibre. To one end of this magnet is attached a lens, and to the other a glass scale adjusted to the lens's focus of parallel rays. This part of the apparatus is inclosed in a box having a small window at each end, on a line with the horizontal axis of the magnet, through one of which light is admitted to the rear of the scale, and through the opposite one the parallel rays from the lens pass out and enter the field-glass of a small telescope, mounted in front, through which the scale divisions may be observed.

This apparatus is mounted on a tripod on which it can be rotated horizontally around the axial line of suspension of the magnet, and the angle of rotation measured on an azimuth scale with vernier attachment, mounted on the tripod underneath the apparatus.

The instrument being adjusted to the proper level, the torsion is first removed from the silk fibre by suspending from it a small plummet of the same weight as the magnet, after which the magnet is suspended so as to come to rest in the magnetic meridian without producing torsion of the fibre, and the movable part of the apparatus rotated till some division of the magnet scale coincides with the cross-wires in the field of the telescope. The reading on the lower scale is then noted, and by a second rotation the instrument is adjusted to the true north, ascertained by observation of one of the heavenly bodies by means of a small transit apparatus mounted back of the magnet-box; and the reading of the lower scale for this second position being noted and the necessary corrections made, the difference of the two readings gives the true declination for the place and time.

CHAPTER V.

ELECTROMAGNETISM.

Deflection by the Electric Current.—Such accidental effects as the magnetizing of steel instruments by lightning had long indicated some relation between magnetism and electricity, but all attempts to produce similar results by artificial means had failed to give satisfactory results. In 1802 Romagnosi of Trente noticed that the magnetic needle was deflected by the voltaic current, but his discovery failed to attract attention. In 1819 Oersted of Copenhagen discovered that the needle was not only deflected by the voltaic current, but tended to take a position at right angles to it, and that the deflection was governed by the direction of the current and relative position of the needle. The discovery, like that of Volta, marks an important epoch in electric progress; it established beyond doubt the mutual relationship of electricity and magnetism, and was the origin of the science of electromagnetism with all the great inventions to which it has given rise.

Oersted's experiments are easily repeated by holding a straightened section of copper wire, connecting the poles of a battery or cell, alternately above and below a poised magnetic needle, and reversing the direction of the current in each position.

The Galvanoscope.—A better instrument for this purpose is represented by Fig. 30, consisting of a mounted, rectangular brass frame, surrounding a poised needle lengthwise, and provided with binding-screws at the terminals of the wires. The under wire has its ter-

72 DYNAMIC ELECTRICITY AND MAGNETISM.

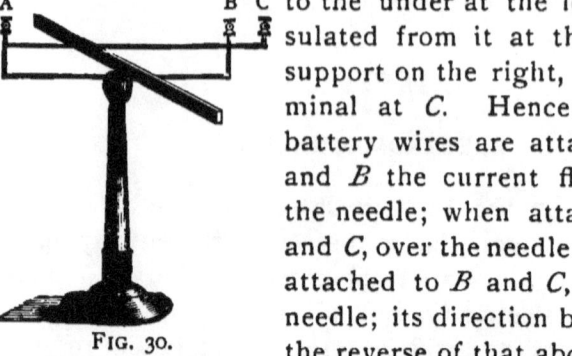

FIG. 30.

minals at A and B, and the upper wire, which is joined to the under at the left and insulated from it at the point of support on the right, has its terminal at C. Hence when the battery wires are attached to A and B the current flows under the needle; when attached to A and C, over the needle; and when attached to B and C, round the needle; its direction below being the reverse of that above.

When the frame is parallel to the needle in the magnetic meridian and the current is flowing over the needle from north to south, the north pole is deflected to the east; when the current is reversed the deflection is to the west; and when the current flows under the needle from north to south or from south to north these deflections are reversed.

When the flow is round the needle lengthwise in either direction, through the connections at B and C, the deflecting force is doubled, the current in the upper and under wires flowing in opposite directions, and hence both tending to deflect in the same direction, as shown above.

Since this instrument may be used to indicate the presence and direction of an electric current, it is known as the *galvanoscope*.

The Schweigger Multiplier.—By multiplying the coils which pass round the needle the deflecting force may be proportionately increased within certain limits. This may be done by winding the wire, provided with an insulating envelope, on a frame of non-magnetic material. The effect may be increased by using a short needle which shall be included within the helix at any angle of deflection, and which, for convenience of observation, may be connected with a light non-magnetic pointer.

The first instrument of such construction was called the *Schweigger multiplier*, in honor of its inventor.

Ampère's Rule.—To determine the direction of the deflection in every case Ampère proposed the conception of a little human figure so placed that the current would enter at its feet and leave at its head, its face being turned constantly towards the needle and its arms extended at right angles to its sides. The left hand would then constantly indicate the direction of the north pole's deflection, and the right that of the south pole, at any point, above, below, or at either side; the deflection in the latter case tending vertically.

The Astatic Needle.—The force which deflects the needle as above acts at right angles to that of the earth's magnetism which tends to maintain it in the plane of the magnetic meridian, and the amount of deflection depends on the relative strength of the electric current, the force of the earth's magnetism at any point being practically constant; but the deflection evidently can never, under these conditions, equal a right angle. But if the effect of the earth's magnetism is neutralized in the apparatus, a much more sensitive and effective instrument can be produced.

This is done approximately by the astatic needle, shown in Fig. 31, which consists of two parallel needles of equal length attached to a short vertical support, with their poles reversed, so that each neutralizes the directive force of the other. They are usually suspended by an untwisted silk fibre so as to be uninfluenced by friction, protected from air currents by a glass case, and

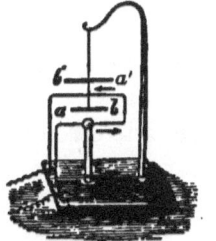

FIG. 31.

provided with a graduated scale to indicate the amount of deflection.

The wire carrying the current passes round one of the needles lengthwise, or may pass round each, being wound alternately in opposite directions; and the poles being also reversed, it is evident that the current flowing in each section of the wire must produce deflection in the same direction in both needles.

But it is practically impossible to construct the two needles with such mathematical precision that there shall not remain a slight deviation in mass, magnetization, and parallelism sufficient to produce a preponderance influenced by the earth's magnetism; so that the very best astatic needles are only approximately correct.

Compensating Magnet.—A similar astatic effect is produced by fixing a magnet in the magnetic meridian with its poles above the similar poles of the needle, and at such a distance that its influence is just sufficient to counteract the directive force of the earth's magnetism. If too close it reverses the needle's position, but with a provision for vertical adjustment it can be maintained in the best position for directive compensation.

Cause of Deflection.—If a copper wire in which an electric current is flowing pass at right angles through a card on which iron filings are dusted, the filings, when the card is lightly tapped, arrange themselves in concentric circles around the wire, indicating that lines of force, due to the current, circulate in planes at right angles to it and magnetize the filings.

If we suppose such a wire to pass through this page, at right angles to the paper, the current to flow from the reader, and a magnetic needle to be carried round it, the needle would constantly tend to assume position in a plane parallel to the paper, its north pole, by Ampère's rule, turning in the same direction as watch-hands move; while if the current flowed toward the reader this direction would be reversed.

From such indications it is evident that there is around every current-bearing wire an electric field in which lines of force circulate in planes at right angles to its length; and since it has already been shown that lines of force in the magnetic field circulate in planes parallel to the magnet's length, it is evident that the tendency of these forces, when brought into mutual proximity, must be to cause the needle or magnet to take position at right angles to the direction of the current, in which position the planes of the magnetic and electric forces coincide.

We may assume a similar physical condition for the electric field to that already assumed for the magnetic, namely, that the air, or the hypothetical ether, is the medium by which the force radiates, and is itself electrified throughout a space of which the current-bearing wire is the central core.

The Electromagnet.—It was discovered in 1820 by both Arago and Davy that iron and steel could be magnetized by the electric current by inclosing a bar of either metal in a helix of insulated wire through which a current is passing; the steel remaining permanently magnetic after being withdrawn from the helix, while the iron is magnetic only while inclosed and during the passage of the current; hence the latter is technically known as the *electromagnet*, in distinction from the former.

Electromagnetic Poles.—By observing the direction in which the current from the battery or other electric generator is passing, the poles may easily be distinguished by Ampère's rule, already given; that being the south pole, viewed endways, around which the current is passing in the same direction as watch-hands move, while that is the north pole around which it flows in the opposite direction.

Winding.—It is immaterial in which direction the helix is wound, whether from right to left, or from left to right, or in layers in either direction, alternately from end to end, like thread on a spool, provided the winding, in each case, is in the same direction throughout; but if reversed, in sections or in alternate layers, the result is consequent poles at the points of sectional reversal, or neutralization between layers oppositely wound.

Magnetic Strength.—An electromagnet is capable of acquiring magnetic strength, as represented by the force in any way exerted, far in excess of the best steel magnet of similar size, and they have been made of sufficient lifting power to sustain more than a ton. The strength is dependent on the size of the iron core, the quality of the iron, the amount of wire in the helix, and the strength of the magnetizing current.

Core.—The *core* can be magnetized only to the point of saturation, beyond which increase in size of helix or strength of current can produce no increase of magnetic strength; hence its mass should be duly proportioned to that of the helix, hollow cores, of sufficient mass, having the same efficiency as solid ones. Its ends should project beyond those of the helix. The iron should be soft and homogeneous in structure to render it capable, not only of the highest degree of magnetization, but of rapidly acquiring or losing its magnetism at the closing or opening of the electric circuit; a quality on which the practical value of the electromagnet largely depends.

Coefficient of Magnetic Induction.—This property of magnetic permeability, or conductivity for the lines of force, is termed the *coefficient of magnetic induction*, and is found in various bodies in different degrees, but in none to the same degree as in soft iron, which is therefore said to have a high coefficient of magnetic induction.

Helix.—The *helix* may consist of fine or of coarse wire of any conductivity, copper being practically the best, and the total volume of current with a given mass of wire may be the same in either case, while the resistance may vary greatly. A helix of ten coils of wire of a given cross-section and length may equal in mass another helix of a hundred coils of one tenth the cross-section and ten times the length: and resistance varying directly as length and inversely as cross-section, and current, with a given E. M. F., varying inversely as resistance, the volume of current in any given section of the fine wire would be only one tenth of that in a coarse wire of equal length, while the total volume in the mass would be the same in either case, since the fine wire has ten times the number of coils.

The rule is to make the resistance of the helix equal to the external resistance of the current, and thus adapt the magnets to the conditions of the work for which each is designed.

The diameter of the coils, within certain limits, does not affect the strength of the magnet, since the field of magnetizing electric force surrounding the wire varies directly in area and inversely in magnetic effect as the square of its distance from the core, variation in one sense compensating variation in the other; since its strength at any point in the larger area, equally distant from the wire, is the same as in the smaller area, while the distance of the wire from the core is proportionally greater.

With such proportion between the mass of the helix and of the core as to insure saturation without excess, the diameter of the helix should not exceed one half its length.

Electromagnetic Saturation.—A perceptible amount of time is required to produce magnetic saturation of the core, which, in the case of very large magnets, may amount to two seconds. It has also been observed that this result is attained more rapidly with high E. M. F. coupled with high resistance than with low E. M. F. and low resistance, though the volume of current in each case is the same.

Form of Electromagnets.—The horseshoe or U form, shown in Fig. 32, in which both poles attract the same armature, has, as in the steel magnet, the greatest practical efficiency. The winding must be in the same direction in both coils, as if a straight bar were thus wound and then bent; which requires the wire to cross to the opposite side at the bend as shown.

FIG. 32.

A modification of this form is seen in the rectangular form shown in Fig. 33; the wire may also be wound on separate bobbins and slipped over the cores as shown.

Armature.—The armature should be of the best soft iron, and of such form and mass as to embrace the greatest practicable number of lines of force, since it is itself a magnet during contact, and its force, whatever it may be, varies in the same ratio as the magnet's force, and the portative force equals the sum of the two. Hence if the magnet's force equals x and the armature's force y, the portative force is $x + y$; and if the magnet's force is

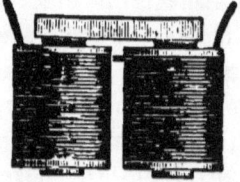

FIG. 33.

doubled, the portative force becomes $2(x+y)$; if halved, $\dfrac{x+y}{2}$; the proportion being the same for any other variation.

The magnetization of the steel magnet can be accomplished most efficiently by the electromagnet.

Experiments in Diamagnetism.—The electromagnet, by its superior power, affords the means of examining diamagnetic bodies, not practicable with the steel magnet. For this purpose Faraday, in 1845, used the apparatus

FIG. 34.

shown in Fig. 34, which consists of two powerful electromagnets, A and B, having hollow cores, between whose opposite poles the body under examination may be suspended, as shown; the distance between the poles being adjusted as required by adjusting the movable frames FF, to which the cores are attached, to the required position, as indicated by the scale RR, where they are secured by the binding-screws EE. The poles terminate in cone-shaped armatures, attached by screws, by which the magnetic force is concentrated between the rounded points.

If the body under examination is repelled from the concentrated magnetic field thus formed, it is classed as *diamagnetic;* if attracted, as *magnetic*, or *paramagnetic*. The tests are made either with small balls of the various substances, suspended near the poles, which are brought into close proximity, or with straight bars suspended between the poles, when placed farther apart. The balls are repelled or attracted as above, according as they are diamagnetic or paramagnetic, while the bar, if diamagnetic, sets itself equatorially, as shown by the bar *ab*, Fig. 34, but, if paramagnetic, axially, that is, parallel to a line joining the poles.

The reason of this becomes obvious when it is considered that the magnetic field, as has been shown, consists of magnetized matter, which may be air at the ordinary density, or rarefied to any extent possible by the formation of a partial vacuum, or some other gaseous body, as oxygen; and the suspended body becomes itself a part of this field when attracted into it. Hence if paramagnetic, like iron, it can form within itself a field which may equal or even greatly exceed in strength that of the gaseous body in which it is suspended, and hence is drawn into that position in which it can embrace the greatest number of lines of force, which, in a straight bar, is the axial position; but if diamagnetic, like bismuth, it is pushed aside by the existing lines of force, from the stronger to the weaker part of the field, where the lines of force are equal to its receptive or magnetic inductive capacity, which in the case of a straight bar is the equatorial position, where the greater part of the bar is most remote from the central region of greatest magnetic intensity. Hence *paramagnetic bodies are those in which magnetic inductive capacity is high, diamagnetic bodies those in which it is low.*

Such a test is not to be regarded as indicating the absolute diamagnetism or paramagnetism of the body under examination, but as a comparison with the magnetic condition of the gaseous medium occupying the field, which would ordinarily be the air at its normal density; the test being analogous in this respect to that for the specific gravity of bodies by a comparison of their weight in air with their weight in water.

Gases are tested by means of bubbles inflated with them showing attraction or repulsion; and liquids similarly, when suspended in glass vessels; but a correction is evidently required in the first case for the matter composing the walls of the bubble, and in the second for the glass of the vessel, since the magnetic condition of either might differ widely from that of the gas or liquid and seriously affect the result.

List of Diamagnetic and Paramagnetic Substances.—The principal substances found to be diamagnetic are as follows : bismuth, phosphorus, antimony, zinc, mercury, lead, silver, copper, gold, water, alcohol, tellurium, selenium, sulphur, thallium, hydrogen, air. The principal ones found to be paramagnetic are as follows: iron, nickel, cobalt, manganese, chromium, cerium, titanium, oxygen; also substances containing the above metals in combination. The proper magnetic classification of platinum is not settled; it has been assigned to each list by different observers, the weight of evidence being in favor of its paramagnetic character; but, when chemically pure, Wiedemann considers it diamagnetic. Flames, smoke, and hot air tend to move, in the magnetic field, from higher to lower potential, which would indicate that they are diamagnetic; but this is not clear, since the movement may be due to the convection of the air and its diamagnetism. It has been observed that bismuth, when pulverized, made into a paste with mucilage, and formed into a roll, sets itself equatorially in the magnetic field, like a bismuth bar; but when compressed into a flat plate its position becomes axial, an effect attributed to the semi-crystalline structure of the mass.

Deflection of the Electric Current by the Magnet.—It has been shown that a magnetic needle or bar magnet free to rotate takes position at right angles to a fixed wire bearing an electric current; conversely it may be shown that if the wire be free to rotate it will take position at right angles to a needle or bar magnet in a fixed position, a result which evidently follows from the law of action and reaction.

Ampère's Table.—This can be explained with the apparatus devised by Ampère, shown in Fig. 35, known as Ampère's *table*, which consists of a wire bent as shown, and suspended so as to rotate horizontally round its centre of gravity; its ends dipping into mercury cups to insure perfect contact; and having arms and supporting standards of brass, with which the battery wires which supply the current connect at bottom, as shown, so that the wire coil becomes part of the circuit.

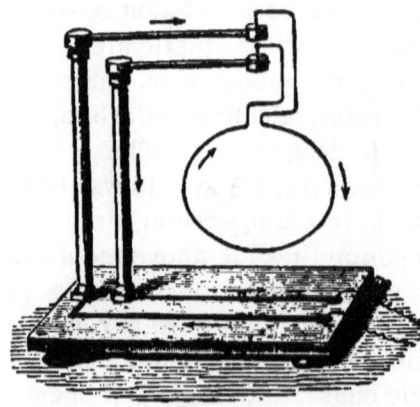

FIG. 35.

This coil will adjust itself with its plane at right angles to the length of a bar magnet thrust into it; its position being reversed when the poles or the current are reversed, and similar effects but weaker being observed when the magnet is held above or beneath the coil.

If the magnet be entirely withdrawn the coil will

assume the same position with reference to the earth's magnetism to that which it would assume if a magnet were placed beneath it with its south-seeking pole turned north, so as to represent the earth's magnetism. The plane of the coil will then be at right angles to the magnetic meridian, and that face turned north which would be indicated by the left hand of Ampère's little figure swimming with the current, face downward, at the bottom of the coil.

The Solenoid.—If this single coil be replaced by the *solenoid* represented by Fig. 36, which consists of a helix with straight portions of the wire returned to the centre as shown, the magnetic effect is greatly increased, each convolution assuming the same position as the single coil, so that the solenoid takes the same position as the magnetic needle and has north and south polarity.

FIG. 36.

This might still be accounted for by the current's reaction setting the planes of the coils at right angles to the magnetic meridian in obedience to the earth's magnetism, but it is found that the poles of the solenoid are repelled by like poles of the magnet and attracted by unlike, which indicates that it has true magnetic properties like those of the needle, which is made further apparent by the fact that two solenoids suspended in mutual proximity behave like two needles similarly placed, exhibiting polar attraction and repulsion in the same manner.

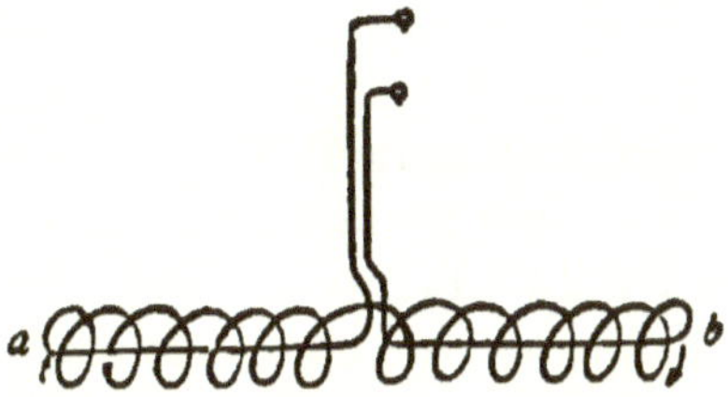

Here, then, we have a current-bearing wire, which may be copper or any other metal, behaving like a steel magnet; which furnishes strong proof, in addition to that already adduced, of the close affinity, if not actual identity, of electricity and magnetism. It is also found that an electric current flowing in rarefied air as its medium, in a glass tube in which the highest attainable vacuum has been produced, is attracted or repelled by a magnet like the current in the wire.

The polarity of the solenoid is much weaker than that of the magnetic needle, but may be greatly reënforced by placing within it a soft-iron core, which in fact makes it an electromagnet. It now takes position in the plane of the magnetic meridian with the energy of the needle, but its polarity must be ascribed chiefly to the magnetism of the core rather than to that of the current, as shown in the solenoid without a core.

De La Rive's Floating Battery.—De La Rive used for the above experiments a little floating battery. It is easily constructed with little plates of zinc and carbon, or copper, attached to a large flat cork and floated on water acidulated with sulphuric acid; a light coil or solenoid attached to the plates projecting from the upper surface of the cork. It is simpler, cheaper, and more easily constructed than Ampère's table, but less effective and limited to a smaller number of experiments.

Mutual Induction of Electric Currents.—By using the rectangular frame represented by Fig. 37 with Ampère's table, the mutual attraction and repulsion of electric currents may be shown. If a current is flowing round the frame, as shown, and a straightened section of another current-bearing wire be held in close proximity, parallel to either of its vertical sections, the frame wire will be *attracted if the two currents flow in the same direction, but repelled if they flow in opposite directions;* this will also be true when the wires are inclined to each other at an angle, *attraction taking*

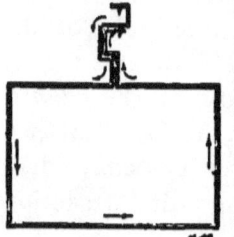
Fig. 37.

ELECTROMAGNETISM.

place when the currents flow either toward or from a common point, but repulsion when one flows toward and the other from a common point.

Ampère, who discovered this mutual action, gave to it the name of *electrodynamics*. Its explanation may be found in the mutual action of solenoids already described, whose like poles were shown to repel and unlike to attract, like the corresponding poles of the magnet. It is evident that when two *like* solenoid poles are brought into mutual proximity, the currents in adjacent sides of their coils must flow in opposite directions, one upward and the other downward; hence the inductive lines of force in the field surrounding the two poles meet in opposition and repel each other as in steel magnets, as already explained. But if *unlike* solenoid poles be brought into proximity the currents in adjacent sides flow in the same direction, and the lines of force interlock and draw the poles together, as in steel magnets. Now if the adjacent sections of the solenoid coils in proximity be straightened, we have exactly the same conditions as in the rectangular wire frame and straightened section of wire in proximity, and the results in each case are the same.

The wires and surrounding air in all these cases are the media through which energy manifests itself by molecular action; and we call these manifestations electric, or magnetic, or electromagnetic, according to the nature of the media, and the conditions under which the manifestation occurs.

When a current flows through a conductor, it is found that the effect of induction is to produce an opposite current in any adjacent parallel conductor.

The nature of this action may be represented by the following diagram:

$(a) + 10 + + + + + + + + + 1$
$(b) - 2 - - - - - - - - - \frac{1}{2}$

86   DYNAMIC ELECTRICITY AND MAGNETISM.

Let a represent a conductor in which a current is flowing from left to right by virtue of the difference of potential represented by $+10$ at the left and $+1$ at the right; and let b represent an adjacent parallel conductor. Since inductive influence radiates from a charged body equally in all directions, only a small fraction of the lines of electric force radiating from a are intercepted by b; but for convenience we may represent this fraction by $-\frac{1}{5}$ at the right, and -2 at the left. Now, since, as has been shown, the positive produces by induction an equal corresponding negative, and *vice versa*, the positive potential in a, represented by $+10$, induces, under the conditions named, a negative potential in b, represented at the adjacent point on the left by -2, while on the right $+1$ in a induces $-\frac{1}{5}$ in b; and since electric movement is always from higher to lower potential, the current induced in b must flow from right to left opposite to that in a.

Hence when currents in two or more adjacent parallel conductors flow in the same direction the effect of their mutual induction is to produce in each a countercurrent, which reduces the volume of the primary current; the effective current by which useful work is accomplished being then represented by the difference between them. As in the illustration, if the primary current were represented by 10 and the induced opposing current by 2, the effective current would be represented by 8.

But if the primary currents flow in opposite directions, the effect of induction is reversed, and the volume of effective current in each conductor increased, as can easily be seen by the following diagram:

$$(c) + 10 + + + + + + + + + 1$$
$$(d) + 1 + + + + + + + + + 10$$

In which the current in c flows from left to right, and in d from right to left. The positive potential 10, at the left of c, induces at the same end of d a much stronger negative than the potential 1 of d can induce in c; consequently the difference of potential between the opposite extremities of d must be greater than if c were removed, and hence its E. M. F. and resulting volume of current must be greater. And the same effect in c must follow from the difference of potential at the right between the 10 of d and the 1 of c.

Rotary Movement by Current Induction.—Fig. 38 represents an apparatus by which the continuous rotary

FIG. 38.

movement of a wire may be produced by current induction. EF represents a circular copper trough containing mercury, round which is wound a coil of insulated coper wire, seen through the opening near m, its terminals being connected with the binding-posts m and o.

From the post n, which is connected with m, a wire leads to the central insulated brass post A, on which is pivoted, in a mercury cup, a wire whose ends, B and C, dip into the mercury in the trough, with which the post p is connected as shown.

The post o is connected with the positive pole of a battery, and p with the negative; the current therefore flows from o through the coil to m, in a direction opposite to that in which watch-hands move, thence through n to the post A, which it ascends, and dividing, passes down the arms B and C, and through the mercury and trough to p.

It is evident, then, that the branch current flowing down C, and that in the coil flowing from o to F, both flow *toward* the common point F, and hence attract each other, while beyond that point the coil current flows *from* F and the current in C *toward* it, and hence repel each other; hence C, being free to move, is impelled *toward* the observer. In like manner it can be shown that B is impelled *from* the observer; so that the wire rotates round the trough in the same direction as watch-hands move, and opposite to the direction of the current in the coil.

By connecting n with o instead of with m, and changing the positive battery wire from o to m, the current in the coil is reversed and the rotation of the wire reversed also.

Water acidulated with sulphuric acid may be substituted for the mercury in the trough, and the lower part of the wire, which is immersed in the fluid, may also be extended by bending it into a circular form so as to conform to the shape of the trough, and furnish a better conductor than the fluid alone.

Rotation may also be produced in mercury or dilute acid, contained in a circular vessel placed above or beneath either pole of a strong magnet, if the vessel be connected with one pole of a battery while a wire from the other pole dips into the center of the fluid, so that the current flows radially either from or toward the center.

A magnet also, loaded so as to float vertically in a vessel of mercury, or pivoted in any other convenient manner, will rotate round its vertical axis, if a battery wire dips into a mercury cup at its upper end while the other pole of the battery connects with the outer edge of the mercury in the vessel.

These experiments may be varied indefinitely, but are all explainable by Ampère's rules for the mutual induction of currents already given; and those with the magnet indicate that the lines of force in the field around its poles are of the same nature as those around a current-bearing wire.

Ampère's Theory of Magnetism.—From observation of various electromagnetic phenomena, such as we have been considering, Ampère deduced the theory that electric currents are in constant circulation around the molecules of all bodies capable of acquiring magnetism, so that every such molecule is the centre of a field of electric force; and that when the body is unmagnetized these currents circulate in different directions, and hence no external effect is produced; but that under the magnetizing influence they are all made to circulate in the same relative direction, and hence between the adjacent sides of any two molecules they are in opposition and neutralize each other; and as this must be true of every interior molecule on all its sides, all interior circulation must cease; while the currents from the exterior sides of the surface molecules, having no opposition, and all now circulating in the same direction round the mass at right angles to its length, produces the effect called magnetism.

In tempered steel the currents retain their harmony of movement after the magnetizing influence is withdrawn, but in soft iron they resume the irregular movement and the magnetism disappears.

This theory, so far as it relates to the external or field currents, is in harmony with the theory of molecular motion already given; while the latter accounts more satisfactorily for the origin of the currents, their permanency in tempered steel and want of permanency in soft iron, by ascribing them to the motion of the molecules themselves, and their change of position under the magnetizing influence. Ampère's theory gives an effect without an adequate cause, while, if molecular motion be regarded as a natural condition of all bodies, which seems to be well established, we have, in its various forms, ample cause not only for those manifestations of energy which we term magnetism and electricity, but also for other kindred physical phenomena.

Generation of Electric Currents by Induction.—In 1831 Faraday made the important discovery that an electric current can be induced in a conductor forming a closed circuit by the movement of a magnet in proximity to it, and also that the same effect can be produced by the similar movement of a current-bearing conductor; both results furnishing additional proof of the close affinity of electric and magnetic induction and of the correctness of Ampère's theory of the circulation of electric currents round the magnet.

Current Induced by Magnet.—These results may be verified in the following manner: Let the terminals of a hollow wire coil be connected with an astatic galvanometer, as shown in Fig. 39; on the insertion of a bar magnet quickly into the interior, the needle will be deflected, showing that an electric current has been induced in the coil. This current is only transient, continuing only during the downward movement of the magnet; but when the magnet is withdrawn the needle is deflected in the opposite direction, showing the induction of a reverse current which continues only during the upward movement. If the poles of the magnet be reversed, the direction of each current will also be reversed in accordance with Ampère's rule.

Similar results may be obtained by placing a soft

FIG. 39.

iron core within the coil, and the alternate approach and withdrawal of each pole of the magnet from its projecting end.

Current Induced by Another Current.—For the current-bearing conductor a small coil connected with a battery may be used as shown in Fig. 40. This is known as the primary coil and fits inside of the other, which is known as the secondary. On its insertion a current is induced in the secondary, the reverse of that in the primary, and on its withdrawal, a direct current, the same as that of the primary, is induced.

If the induced effect be regarded as magnetic, then in the experiment with the magnet it is evident that on the insertion of either pole the opposite polarity is induced in the coil, so that the induced current flows in the reverse order to that which would produce the existing polarity of the bar, but on its withdrawal the opposite effect is produced, and the induced current is direct, the same as would produce the existing polarity of the bar.

FIG. 40.

The same would be true of the experiment with the primary coil, which may be regarded as another form of bar magnet with currents circulating in its coils in a similar manner to those which, as shown by Ampère, circulate in the field of the magnet; hence we have the inverse current on insertion, as has been shown, and the direct current on withdrawal.

On the insertion of either the magnet or primary coil a varying number of lines of force is cut by the secondary coil, and this creates a difference of potential between that part of the secondary coil which cuts the lines and that which does not, which becomes an important factor in the generation of the current; such generation being always a result of difference of potential.

It will also be noticed that the inverse current is induced when the movement is such as to produce increase in the number of lines cut, and the direct current when it is such as to produce decrease; the force of the induced current being always in such direction as to oppose the mechanical movement by which it is produced, according to what is known as *Lenz's law*.

Current Induced by Opening or Closing Primary Circuit.—If a primary coil be placed inside of a secondary and insulated from it, electric currents can be induced in the secondary circuit by opening or closing the primary circuit. Let the apparatus be arranged as shown in Fig. 41; the secondary coil connected with the galvanometer

FIG. 41.

G, and the primary with the battery P; the latter connection being made, for convenience, through the mercury cups $g\ g'$. No movement of the needle occurs while the connections remain undisturbed, but if the primary circuit be opened by lifting a wire terminal from one of the mercury cups a deflection of the needle occurs, indicating a transient current in the secondary circuit in the same direction as that in the primary; on closing the primary circuit an opposite deflection of the needle indicates a transient current in the secondary circuit the reverse of that in the primary. By an experiment similar to this, Faraday made the discovery of current induction as stated.

The terms "*make*" and "*break*" are used for convenience, to denote respectively the opening and closing of a circuit in this manner.

Current Induced by Varying the Strength of Primary Current.—If a short circuit be arranged by a wire making direct connection between the mercury cups, by which a portion of the current can be diverted from the primary coil, then, on closing this short circuit, the primary current is weakened and a transient, direct current induced in the secondary coil ; but on opening it, so that the full current again flows through the primary coil, a transient reverse current is induced in the secondary coil.

Here also it is seen that the inverse current results from increase of inductive influence, and the direct from decrease.

Results of Current Induction.—These results may be summarized as follows : 1. *A magnet, moving in close proximity to a conductor forming a closed circuit, induces in it an electric current which continues during the movement, and is the reverse of that which would have produced the magnet's polarity when the movement is such as to increase the inductive influence, but the same as would have produced the magnet's polarity when the movement is such as to decrease the inductive influence.*

2. *A current-bearing conductor, moving in close proximity to a conductor forming a closed circuit, induces in it an electric current, which continues during the movement, and is the reverse of the primary current when the movement is such as to increase the inductive influence, but in the same direction as the primary current when the movement is such as to decrease the inductive influence.*

3. *A current-bearing conductor, placed in close proximity to a conductor forming a closed circuit, induces in it, when the primary circuit is opened, a transient electric current in the same direction as the primary current, and an inverse transient current when the primary circuit is closed.*

4. *A current-bearing conductor, placed in close proximity to a conductor forming a closed circuit, induces in it, when the primary circuit is weakened, a transient electric current in the same direction as the primary current, and an inverse transient current when the primary current is strengthened.*

Generation of Current Dependent on Variation of Intercepted Magnetic Force.—The E. M. F. generated by these mechanical movements varies as the number of lines of force cut by the conductor *per unit of time*, and the strength of the induced current varies as the E. M. F. divided by the resistance of the circuit. But there must always be a difference in the number of lines of force cut by different parts of the conductor, whether it be moved with reference to the field or the field with reference to it, in order to produce difference of potential. No current can be induced in a conductor moving parallel to itself in a perfectly uniform field, since there is no difference of potential ; but if the field vary in uniformity, or different parts of the moving conductor, in a uniform field, cut a varying number of lines of force, there is, in either case, generation of current as a consequence of difference of potential. This development of the original discovery is also due to Faraday, and its application in the construction of electromagnetic apparatus for practical use is of the highest importance.

The transient electric currents can, by various devices, be produced in such rapid succession as to be practically continuous, and either direct or alternating as required.

Coefficient of Mutual Induction.—The inductive energy of a current-bearing conductor is found to vary as the area which it is capable of inclosing and as the strength of the current which it carries ; and the mutual inductive action of such conductors on each other, as represented by the number of lines of force intercepted in each by the other, is known as the *coefficient of mutual induction;* and this quantity must attain its relative maximum in such conductors when each intercepts the greatest possible number of lines of force in the others, which must evidently be when they are in the closest relative position. But its absolute value, with unit current, must vary as the total area capable of being inclosed by the conductors which, in a coil, is represented by the number and diameter of the convolutions.

But since mutual induction is only one of the conditions of the generation of E. M. F. by the coil, it must not be understood that such generation varies as this quantity alone, but is dependent on a concurrence of the conditions already mentioned.

Self-Induction.—A current-bearing conductor having the form of a coil, or any form in which different parts of the same circuit are brought into mutual proximity, experiences a momentary inductive effect on the opening or closing of the circuit similar to that produced in a secondary coil.

This, as in the secondary coil, induces a transient current in the same direction as the original current on opening the circuit, and in the opposite direction on closing, which is known as the *extra* current. In a straight conductor, no part of which returns so as to be adjacent to another part of itself, there is still a slight self-induction, which, however, is of little practical consequence; but in coils, the close proximity of the numerous turns greatly increases it. The slight self-induction noticeable in a straight conductor may result from its being produced in an interior shell, as represented in cross-section, by the original current in the outer shell.

Extra Current.—It is evident that the inverse extra current, induced on closing the circuit, must neutralize the original current to the extent of its strength, while the direct, induced on opening, adds its strength to the original; hence it is found that when a primary coil is inclosed within a secondary, the transient current induced in the secondary on opening the primary circuit is much stronger than that induced on closing it. The extra current at "make" retards the maximum effect for an infinitesimal fraction of a second, so that the transient current induced in the secondary does not attain its maximum till just before it ceases; and in like manner, at "break," the maximum is not attained till an instant after the break occurs, and just before the transient current disappears.

The case is analogous to the flow of water in a trough under the influence of gravity; the inertia of the mass retarding the current when it begins, but continuing it, with accumulation of water in the trough, when suddenly interrupted.

The Spark.—The high E. M. F. of the primary current at "break" produces an electric spark between the terminals of the secondary coil when separated by an airspace, which varies from a fraction of an inch to several inches, and is similar to that which occurs between the poles of a static electric machine. No spark occurs at "make" except in very large coils, and even in these it is feeble, indicating clearly a much lower E. M. F. than at "break," as already explained.

Induction of Core.—It has been shown that the magnetic energy of a solenoid is greatly increased by the insertion of a soft-iron core, and reciprocally, that in the electromagnet the core acquires its magnetism from the electric current in the inclosing coil. The core, in either case, forms a medium of high conductivity or magnetic permeability greatly superior to that of the air which it displaces, and its magnetism reacts as a

98   DYNAMIC ELECTRICITY AND MAGNETISM.

coefficient of the magnetism and electricity of the inclosing coil; so that the magnetic and electric strength developed is in proportion to the *coefficient of magnetism* in the core, as this property of permeability has been termed.

Induction-Coil.—The principles of current-induction thus developed have given rise to the apparatus known as the *induction-coil*, or inductorium as it is also called, the construction of which, as improved, is due to Ruhmkorff.

FIG. 42.

Its general construction, with some variation in minor details, will be understood from Fig. 42. It consists of a primary coil of a few layers of cotton-wound, coarse copper wire, usually No. 14-20, inclosing a core made of a bundle of soft-iron wires. This coil is inclosed in a secondary coil of very fine, silk-wound copper wire, usually No. 26-30, several hundred times the length of the primary, from which it is insulated by a hard-rubber tube. On account of the extreme fineness of this wire, great care is required in the winding, to have the layers perfectly even, and to avoid a cross or a break. A coating of paraffine is applied during the winding to improve the insulation, and the layers are separated by paraffined paper.

Condenser.—The coil thus completed is mounted on a base of wood or hard rubber, in the bottom of which is placed a *condenser* consisting of a number of sheets of tin-foil insulated from each other by paraffined paper, the alternate ends of the foil projecting beyond the paper at each end, so that all the oddly numbered sheets are in contact at one end, and all the evenly numbered sheets at the other end ; and each end is connected with an interior terminal of the primary coil, the exterior terminals of which are connected with the battery; so that the condenser is directly in the primary circuit. It is sometimes omitted from small coils.

Interrupter.—An *interrupter*, or contact-breaker, is also placed in the primary circuit, mounted on one end of the base, and varies in construction according to the size and design of the coil. In small coils it consists of a light steel spring known as the *vibrator*, carrying at one extremity a small armature placed opposite the projecting end of the core, which closes the circuit by the pressure of a platinum plate, attached to it, against a platinum point fixed in the end of a set-screw; but on the passage of the current the magnetism of the core attracts the armature and opens the circuit; the cessation of the current demagnetizes the core, and releases the armature, which, being forced back by the spring, again closes the circuit as before. The set-screw regulates the amplitude of the vibration, and a second set-screw is sometimes used to regulate the tension of the spring.

The make and break thus produced occur with great rapidity, giving rise to a series of transient, alternating currents, induced in the secondary coil, which thus become continuous, and are known as the *faradic* current. If a weak faradic current is desired, the amplitude of the vibrations is reduced, and if necessary the tension of the spring also, so that the primary circuit is opened or closed before the transient current in each case has attained its maximum; but if a strong faradic current is desired, the amplitude of the vibrations and tension of the spring are increased and the opposite effect produced.

The vibrator can be used for coils giving sparks 17 to 28 inches in length, but in very large coils the extra current at break produces a series of sparks between the points of contact which are liable to melt the platinum point, injure the coil by heating, and interfere with the promptness of the break by their conductivity; hence interrupters not liable to this defect are required. The one shown in Fig. 42 is Foucault's, and consists of a brass lever L, supported on a vertical spring, its left end carrying an armature, shown just above the projecting end of the core, and its right end a plunger tipped with platinum, which dips into the mercury in the cup M, the surface of which is covered with alcohol. The attraction of the armature lifts the point out of the mercury, and the alcohol, being a non-conductor, prevents the spark, the contact being renewed again by the force of the spring. The adjusting screws shown regulate the amplitude of the movement for currents of different strength.

Interrupters operated by clock-work, or by some external force, are also used; and when sparks alone are required at intervals, they may be operated by hand. One used by Spottiswoode consists of a brass wheel having a number of radial slots filled with hard-rubber, and rotated by a little engine ; a platinum spring presses on the circumference, opening the circuit when in contact with the rubber, and closing it when in contact with the brass. By this means the make and break can be produced with great rapidity, and the smoothness of the induced current proportionately increased.

Sliding Core.—It is often desirable, especially in coils for medical use, to vary the strength of the induced current to a greater extent than can be done by varying the amplitude of movement in the interrupter ; this may be done by varying the magnetism in the primary coil by sliding the core in or out to any extent required for the variation. In such case a small electromagnet, placed in the primary circuit, is used to operate the interrupter instead of the magnetism of the core.

Water Rheostat.—The same object may be accomplished by varying the resistance of the primary circuit; which can be done by connecting it with a column of water contained in a vertical glass tube set on the base; one terminal being let into the tube at bottom, while the other is attached at top to a plunger, by which the distance between the terminals, and consequently the resistance, can be varied as desired ; the plunger having sufficient sliding friction to retain it at any point. The term *rheostat* is used to designate any apparatus thus used for resistance, this particular kind being known as the *water rheostat*.

Construction of Core.—A bundle of wires is preferred to a solid bar for the core, since, as in laminated magnets, it prevents the formation of Foucault currents by giving the lines of force a normal direction toward the poles. They should be bound together by soldering the ends or otherwise, so as to be moved as one mass, and to furnish an even surface at the end next the interrupter for its proper adjustment.

Operation of Condenser.—The object of the condenser is to absorb the extra current which occurs at "break," and thus prevent the spark which is liable to occur at the interrupter, and which interferes with the suddenness of the break by its conductivity; this absorbed current opposes the reverse current induced at "make" and retards its maximum effect. Hence the effect of the condenser is an increased promptness of action at "break," and a decrease in promptness, or partial suppression, at "make," so that the induced current at "break" becomes practically the current of efficient work.

Leyden Jar as a Condenser.—An insulated Leyden jar or Leyden battery may be used as a condenser in the secondary circuit, to increase the energy of the spark, by connecting its coatings respectively with the opposite terminals; knobs from the opposite coatings being brought within sparking distance, a series of sparks of greatly increased energy passes between them, similar to those which pass between the opposite poles of a Holtz or a Töpler machine; a result which is proof of the identity of static and dynamic electricity.

Special Construction.—In large coils special methods of winding and insulation are required on account of the high induction and potential difference between the primary and secondary coils and between different parts of the secondary. This is necessary to prevent short-circuiting and permanent injury by burning, in case a spark, from insufficient insulation, should cross between the coils or between the layers of the secondary; also to bring the coils into the best inductive relation, and to reduce what is known as the "Leyden jar effect" which occurs between the outer coating of the primary and inner coating of the secondary, and interferes with electric action.

It is found that in the primary coil induction is greatest at the centre and least at the ends; also that the potential difference in the secondary is least at the centre and greatest at the ends ; hence in large coils the insulation between the primary and secondary, which should always be ample, is often adapted to these con-

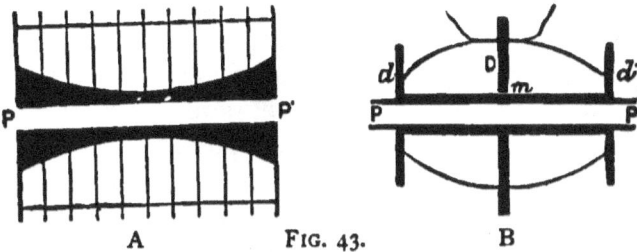

A   FIG. 43.   B

ditions by making the insulating tube thicker at the ends than at the centre, as shown at A, Fig. 43. This gives increased insulation at the ends where it is most required, and brings the greater mass of the secondary coil to the centre where induction is the strongest.

The same results may be accomplished in a different manner, as shown at B, by making the insulating tube of the same diameter throughout, with projecting ends for greater end insulation, and winding the secondary coil in an elliptical form with the greater mass of wire at the centre. The primary coil in both cases occupies the interior of the tube.

It is evident that the potential difference must be much greater between adjacent layers, especially at the ends, than between adjacent convolutions, since each convolution adds its quota to this difference, and each layer being doubled back on the one next underneath, the convolutions of each pair thus formed are separated at each alternate end from those of the next adjacent pair

104 DYNAMIC ELECTRICITY AND MAGNETISM.

by the whole number of convolutions in the two layers. This produces, in large coils which require long layers, too great a strain on the insulation at those points; hence such coils are usually wound in two or more sections, as shown at B, insulated from each other by hard-rubber, through which the wire passes to connect them, the layers being thus shortened and the strain reduced.

Very large coils have these sections subdivided into thin disks, separated also by hard-rubber, through which the wire passes as above. The wire of different sections may also vary slightly in diameter, the coarser wire being used for the end sections and the finer for the middle, on account of the greater potential difference at the ends and induction at the center already mentioned.

Ruhmkorff's Commutator.—The commutator invented by Ruhmkorff is often used in connection with the coil, either to reverse or to interrupt the current. Its construction will be understood from Fig. 44. A hard-rubber cylinder mounted on a base has two brass cheeks V and

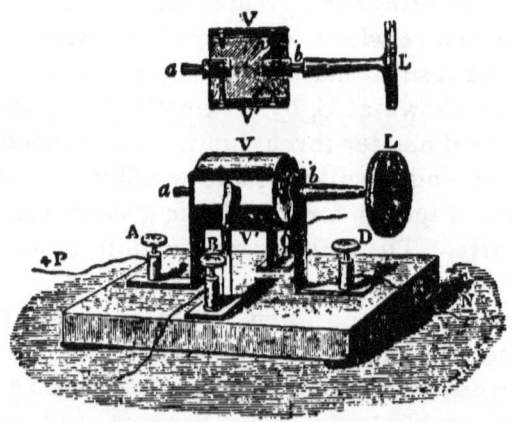

FIG. 44.

V' connected by the pins v and v' respectively with the axes a and b, which, as well as the cheeks, are insulated from each other by the rubber. Two brass springs connected with the binding-posts B and C press against the cylinder, which is mounted on brass supports connected with the binding-posts A and D. The battery wires connect also with A and D, and the terminals of the primary coil with B and C.

When the cylinder is in the position shown in the cut, with the springs pressing against the insulating rubber, no current can pass; but when turned so as to bring the cheek V in contact with the spring attached to B, and V' with that attached to C, the current flows from $+P$ through $A a v' V'$, through the coil from C to B, and thence through $V v b D$ to $-N$. But when the cylinder is reversed, so that V connects with C, and V' with B, then the current from $+P$ to $-N$ flows through the coil in reverse order, by way of $A a v' V'$, from B to C, and thence, as before, through $V v b D$ to $-N$.

The Coil a Converter.—The coil is not a generator but a converter, transforming the energy derived from the battery, or any dynamic generator, by increasing the potential difference, or E. M. F.; which must be done at the expense of a corresponding reduction in the volume of current, since otherwise there would be an increase of electric energy without a corresponding expenditure of chemical or other energy, which would be impossible. For all the energy is derived from the generator, and a certain percentage expended in operating the interrupter and overcoming the resistance of the primary coil, so that even when the interrupter is operated by external power there is still a loss from the resistance of the primary.

This energy, as has been shown, first enters the primary coil, which, from its low resistance, carries a large current, while its high coefficient of magnetic induction, derived from the core, multiplies the lines of force cut by the secondary coil. The secondary, from the extreme fineness of the wire, has great resistance, and hence carries a very small current, but creates a great E. M. F., or potential difference, which in a large coil may equal many thousand volts, the convolutions being very numerous and each adding its quota to the coefficient of mutual induction.

But since, with wire of any given cross-section, the resistance increases directly as the length, and since, as shown, the E. M. F. also increases as the length and hence in the same ratio, the current must remain constant. But since, with a given size of coil, any variation in cross-section of wire produces a corresponding opposite variation in its length, from which must result a corresponding variation in the relative proportions of E. M. F. and current, it is evident that any resulting increase of current must produce a decrease of E. M. F., and any decrease of current an increase of E. M. F.

The length of the spark, or discharge, depends both on the E. M. F., or electric pressure, and on the cross-section of the perforation made through air or other insulating medium; for the length and cross-section of the perforation measure the resistance overcome, and any variation in either dimension must be compensated by an opposite variation in the other, otherwise there would result an increase of work without a corresponding increase of energy. Hence the great spark-length obtained by the coil is the result of the transformation, which, by reduction in volume of current, concentrates the electric energy on a fine line and impels it with a corresponding increase of E. M. F. The great advantage of the coil in this respect is shown by the fact that the longest spark obtainable without a coil from 1080 silver chloride cells, the largest battery ever constructed, is only $\frac{1}{250}$ of an inch, while Spottiswoode's great coil gives, with 30 Grove cells, a spark of 42½ inches in length.

Electric Perforation.—Perforation, as used above, refers to the path by which the electric energy passes through a substance, using its material as the medium of transfer; displacement of this material being often an accompaniment of the discharge, though not a necessary consequence, since energy and not matter is thus transferred. Paper, for instance, when thus perforated is displaced, while glass is pulverized on the line of discharge, with surrounding fracture and little or no displacement. A discharge through any insulating medium is termed disruptive.

Physiological Effects of Faradic Current.—The faradic, or alternating, current of the coil, when passed through any part of a living body, produces a tingling sensation accompanied with muscular contraction, which may be mild or painfully severe, according to the strength of the current, as regulated in the manner already described. This current is now extensively employed in medical practice, and its use constitutes an important branch of electro-medical treatment.

In ordinary lecture-room experiments it is received through metal handles, connected with the coil terminals, which may be held in the hands or otherwise applied; but, for medical use, special electrodes, such as sponges and rollers having insulating handles, are used, by which the current can be applied by the physician or attendant as required.

Discharge in Air and in Vacuo.—The intensely brilliant spark produced by the electric discharge in air at the ordinary density is due to the heat generated by the electric energy in this high resisting medium, which is rendered incandescent on the line of discharge by the intensely rapid vibration of its molecules. Hence *the* *electric spark is a line, or fine cylinder, of incandescent air, often bent, contorted, or subdivided, whose molecules are in a state of intensely rapid vibration.*

The longest spark in air at the ordinary density is comparatively short, the energy being soon expended in overcoming the high resistance; but when the density is reduced by the production of a partial vacuum, the length of the discharge is proportionally increased. This may be done by a partial exhaustion of the air from a glass tube with the common air-pump, but is accomplished more effectively in the hermetically sealed vacuum tubes of Geissler, in which the density is reduced by the mercury pump to $\frac{3}{1000}$ of an atmosphere, and platinum terminals sealed into the extremities. A discharge several feet in length may be obtained in such a tube with a small coil; the low resistance also permitting increase of cross-section, with change of color to the light pink seen in the aurora, which is a similarly diffused discharge. This change of color is a necessary consequence of the diffusion, since enlargement in the space occupied by the discharge produces a corresponding diffusion of the light and heat produced at each point and hence a proportional reduction of their intensity.

Since the space occupied by the discharge increases in the same ratio as the reduction of the atmospheric density, a reduction to $\frac{3}{1000}$ of an atmosphere would give, with a tube of sufficient size, an enlargement of $333\frac{1}{3}$ times the space occupied by the same discharge in air at the ordinary density.

But when the density is reduced to $\frac{1}{1000000}$ of an atmosphere, as in Crooke's vacuum tubes, the medium becomes insufficient to carry the current with the same facility as in the lower vacuum, and we have resistance in the opposite sense to that found in air at the ordinary density, with many interesting phenomena described in "Elements of Static Electricity," in which this subject, and also the auroral discharge, is more fully discussed.

Electric Gas Lighting.—The coil and battery are extensively used for lighting the gas in churches and audience halls where the burners are not easily accessible. For this purpose wire, properly insulated, is connected with the chandeliers, and interrupted at each burner so as to furnish short sparks which pass in series through the escaping gas and light it.

Spark Coil.—The *spark coil*, as it is termed, is best adapted to this use; it consists of the primary coil and core, giving a short thick spark with strong current; the secondary coil and interrupter being dispensed with, reducing the resistance, risk of burning out, and expense.

The term *spark coil* is also applied to the complete induction coil, when constructed for this or any similar purpose, where the main object is the spark rather than the current.

The induction or influence machine—Holtz, Töpler, or Wimhurst—is also used for the same purpose, as described in "Elements of Static Electricity."

CHAPTER VI.

ELECTRIC MEASUREMENT.

ELECTRIC MEASUREMENT pertains to measurement of the force exerted in any way by electric energy, or of the resistance which opposes it, or of certain effects resulting from the mutual relations of this energy and resistance. It is dependent on certain physical conditions which will now be considered in their order.

Electric Potential.—Potential in the physical sense is that condition of matter by virtue of which it is capable of exerting physical force. Thus we estimate the heat potential of a body by the effect it can produce on temperature; its gravity potential by the attractive force it can exert as a mass; its magnetic potential by the magnetic force it can exert; and its electric potential by the electric force it can exert.

Electric potential is designated relatively as positive, negative, or zero. Matter has *positive* electric potential, or is positively electrified, when its electric condition is higher than that of other matter to which it may be related either by contiguity of position or electric connection, so that it is capable of imparting electricity to it; it has *negative* electric potential when its electric condition is lower, so that it is capable of receiving electricity from such other matter; and *zero* electric potential when its electric condition is the same as that of the other matter, so that it can neither impart nor receive: and any variation in either condition must of course change these relative electric conditions. Hence a body may have positive potential with reference to one of lower potential and, at the same time, negative with reference to one of higher potential, or zero with reference to one of the same potential.

Potential difference is conveniently represented by the symbol *p. d.*

Electromotive Force.—Electromotive force has been already briefly referred to as "that which moves or tends to move electricity from one point to another," and as being represented by difference of electric potential. Hence it is the relative condition of electric force between different bodies or parts of a body; the tendency of electricity being always to move from higher to lower potential in the same sense as heat tends to move from higher to lower temperature. It is sometimes represented as *electric pressure*, in the same sense as water pressure in a reservoir, or steam pressure in a boiler, and the analogy is correct so far as the *pressure* is concerned; but water or steam pressure tends to move matter, while electric pressure tends to move molecular force, using matter as its medium.

It is independent of the quantity of electricity generated, and depends solely on potential difference, just as force in each infinitesimal drop of water in Niagara Falls is derived from the height of the falls and not from association with other drops. Hence small electric quantity may be combined with large electromotive force, or the reverse. The number of battery cells joined in parallel may be multiplied indefinitely, while the electromotive force, as has been shown, is only that of a single cell, each cell being, in this respect, independent of the others; and the same is true of the generating parts of any other electric generator, when joined in parallel, as the parallel pairs of plates in a Töpler machine, or the parallel coils of a dynamo. But when these parts are joined in series, the electromotive force of each being added to that of the others varies as the number of such parts and as the value of this quantity in each.

The symbol of electromotive force is E. M. F., as already given, but in mathematical formulæ E alone is used.

Electric Resistance.—Electric resistance, as briefly defined in Chapter I, is that which opposes electric movement, and may consist specifically in the molecular constitution of the conductor or insulator; in counter-electromotive force, or counter-induction; in useful work; or in an artificial obstruction placed in the circuit for a useful purpose.

In conductors it varies directly as the length of the conductor and inversely as its cross ssction, and also inversely as its conductivity; and as insulators must be regarded as inferior conductors, the same rule applies to them. But the relative electric resistance of the substance displaced by the insulator must also be considered, since an insulator, as glass or vulcanite, required in construction, may displace dry air which has much higher electric resistance than either, and the insulation be thereby reduced. In such case resistance is increased by reduction in the cross-section of the insulator. But if a substance of lower resistance is displaced, increase in cross-section of the insulator increases the electric resistance.

In its effect, electric resistance in a conductor is similar to the frictional resistance produced in a pipe conveying a fluid, by an accumulation of loose material, such as moss or cotton waste, which obstructs the flow; but in its nature it is very different, fluid matter being transmitted by the pipe, but electric energy by the conductor.

Insulation and Conductivity.—Resistance is the opposite of conductivity, and very high resistance, when applied to a certain class of bodies, is termed *insulation*. Conductivity is that quality of a body which facilitates electric transmission, while resistance or insulation obstructs it. Each varies inversely as the other, but there is no well-defined boundary between them; every conductor having a certain amount of resistance, and every insulator a certain amount of conductivity. Where conductivity is found to predominate, as in the metals, the term *conductor* is applied, and where resistance predominates, as in glass and vulcanite, the term *insulator* is applied. Silver and copper are metals of the highest conductivity and consequently of the lowest resistance. German-silver and bismuth have high resistance and hence low conductivity. Glass and vulcanite have high insulation and correspondingly low conductivity, so low that the term conductor is never applied to them, nor is the term insulator ever applied to silver or copper. Hence a conductor is any substance of such low resistance that it can be used practically for the transmission of electricity, and a non-conductor or insulator is any substance of such high resistance that it can be used practically to prevent such transmission.

If electricity is a mode of molecular motion by which energy manifests itself, difference of molecular constitution in different bodies would easily account for these varied results. Such difference might consist in variation in the size, shape, or relative arrangement of the molecules, or in a combination of such causes. Molecular arrangement in a conductor might be such as to produce harmony of movement by which undulations would be rapidly propagated, while in an insulator a different arrangement might produce conflicting movements, by which they would neutralize each other and thus prevent transmission, as already explained in regard to magnetism in Chapter IV.

Electric Current.—Current, as stated in Chapter I, is that electric condition in a conductor which results from electromotive force modified by resistance; and its mathematical quantity is ascertained by dividing the former by the latter. It pertains exclusively to what is understood as electric movement, and is used in the same sense when applied to this movement in a conductor, as the same term when applied to the flow of water, steam, gas, or any other fluid, in a pipe; and in this sense also are used the terms current intensity, quantity, volume, strength and resistance; and on this principle all the various kinds of electric apparatus pertaining to current are constructed, and current estimates and measurements made.

In the present imperfect state of electric knowledge this conventional form of expression is convenient and admissible, provided the distinction between an electric current and a fluid current is kept strictly in mind, the former being a flow of energy, the latter a flow of matter. Our actual knowledge of the nature of an electric current is very limited; the generator creates E. M. F. at one end of the conductor, and a molecular movement is supposed to take place by which electric energy is instantly transmitted. Theoretically this movement is in the form of transverse vibrations, but as a matter of fact its nature is unknown; the only well established fact concerning it being that there is no transmission of a fluid, as was formerly supposed, or of other matter, energy alone being transmitted, using matter as its medium; and it is the effect produced by the energy on this medium, which is known as the electric current, or electricity in process of transmission.

Ohm's Law.—The law by which the strength of an

electric current is determined was discovered by the German electrician, Ohm, and is briefly as follows:

The strength of an electric current varies directly as the electromotive force by which the current is impelled, and inversely as the total resistance encountered.

From this law are derived the following formulæ by which either of the three factors represented by the symbols C, E, R can be found when the other two are known:

Formula for finding current, $C = \dfrac{E}{R}$.

Formula for finding E. M. F., $E = CR$.

Formula for finding resistance, $R = \dfrac{E}{C}$.

Electric Units.—In order to render possible the calculations required in estimating electromotive force, resistance, and current, certain units of measurement are required, some of which have already been briefly defined in connection with batteries. They are appropriately named after different distinguished electricians.

The International Electric Congress which met at Paris in 1881, and again in 1884, revised these units, giving them a definite value referable to fixed standards, and the units thus established are distinguished as "*legal*" and accepted as authoritative.

The C. G. S. mechanical unit is taken as the basis of the units by which electric force may be represented in absolute measure. The initial letters, C. G. S., are the symbols of the three factors, space, mass, and time; C. standing for centimeter, G. for gramme, and S. for second; hence this C. G. S. unit represents the work accomplished by the movement of a mass equal to one gramme, through a space equal to one centimeter, in one second, and is known as the *erg*.

116 *DYNAMIC ELECTRICITY AND MAGNETISM.*

The Volt.—The volt is the unit of electromotive force, and was formerly represented by the E. M. F. of a battery-cell nearly equal to that of the Daniell; but as this is a variable quantity, a definite value was given this unit by the adoption of a quantity represented by 100,000,000 C. G. S. units as its equivalent, which is therefore the amount of electric energy which, if converted without loss, would equal this amount of mechanical force. But as such a large number is inconvenient to write, the equivalent expression, 10^8, has been adopted in its stead, and the same method of abbreviation followed in representing the other electric units. Hence the legal volt equals 10^8 C. G. S. units of E. M. F.; but in approximate estimates, the E. M. F. of a Daniell cell, which is about 1.05 volts, is usually sufficiently accurate. The *microvolt* equals $\frac{1}{1000000}$ of a volt.

The Ohm.—The unit of electric resistance is the *ohm*. It was formerly represented by the resistance of a given number of feet of wire of a given gauge; a very unreliable standard, requiring a different length for each different metal, and subject to great variation from difference of quality or temperature, or slight difference of gauge.

The standard resistance adopted by the Electric Congress is that of a column of pure mercury, 106 centims. in length, and 1 sq. millim. in cross-section, at the temperature of 0° C.; which is nearly equal to 10^9 C. G. S. units; the resistance of a similar column, 106.21 centims. in length, being the exact equivalent, but to avoid the fraction the standard was fixed as above. Hence the legal ohm equals 10^9 C. G. S. units of resistance. The *megohm* equals a million ohms.

The Ampere.—The unit of current strength, or volume, is the *ampere*, and is derived from the two preceding units in accordance with Ohm's law, by dividing the unit of E. M. F. by the unit of resistance. Hence, since

$$C = \frac{E}{R}, \quad \text{1 amp.} = \frac{10^8}{10^9} = 10^{-1} = \frac{1}{10}.$$

Hence the legal ampere represents current strength equal to $\frac{1}{10}$ of a C. G. S. unit. It does not include time as an element, but refers exclusively to the strength of current flowing in a conductor at any instant, as represented in cross-section at any point. The *milli-ampere* is the thousandth part of an ampere.

The Ampere-Hour.—The ampere-hour is a unit derived from the last, in which the element of time is included. It represents a current of one ampere flowing through a conductor for one hour, or its equivalent in a greater current for a less time or a less current for a greater time, as two amperes for half an hour, or half an ampere for two hours. It is of recent origin, but is sanctioned by general use, and is often convenient in electric calcualtions.

The Coulomb.—The unit of current quantity with reference to time is the *coulomb*. It is derived from the ampere, and represents the quantity of electricity which flows for one second with a current strength of one ampere. Hence any variation either in the time or strength of a current produces a corresponding variation in the quantity represented in coulombs, while if one factor varies inversely as the other the quantity remains constant; a ten-ampere current flowing for one second or a one-ampere current flowing for ten seconds represents ten coulombs. And since there are 3600 seconds in an hour, 3600 coulombs equal one ampere-hour. The legal coulomb, being derived from the ampere, equals 10^{-1}, or $\frac{1}{10}$, of a C. G. S. unit of current quantity.

The Farad.—The electric unit of capacity is the *farad*. It represents the storage of one coulomb of electricity in a condenser; and when such storage raises the potential to one volt, the capacity equals one farad. The legal farad equals 10^{-9}, or $\frac{1}{1000000000}$, of a C. G. S. unit of capacity.

The Microfarad.—The farad being inconveniently large for practical use in estimating the capacity of condensers, the microfarad, representing one millionth of a farad, has been adopted in its stead. Hence the microfarad equals 10^{-15}, or $\frac{1}{1000000000000000}$, of a C. G. S. unit of capacity.

The Watt.—The unit of electric power is the *watt*, named after the inventor Watt. It is derived from E. M. F. and current combined, neither of which taken alone is a correct representative of electric power; E. M. F. representing pressure, while current represents pressure modified by resistance; hence there might be large E. M. F. with small power, or the reverse, in proportion to the relative resistance; or current might remain constant while power varied. Hence, to obtain an accurate expression for electric power, or *rate of work*, the E. M. F. is multiplied into the current,—that is, the volt into the ampere. The legal watt then equals one volt multiplied into one ampere, the product being 10^7 C. G. S. units of power,—$10^8 \times 10^{-1} = 10^7$. The term *volt-ampere* is synonymous with *watt*.

The Electric Horse-Power.—The electric horse-power, which is the equivalent of the mechanical horse-power, is represented by 746 watts, equal to

$$746 \times 10^7 = 7,460,000,000 \text{ C. G. S. units of power.}$$

Different Kinds of Electric Measurement.—The electric measurement here considered pertains to dynamic electricity; and since much of the apparatus by which electricity in this form is generated, and by which it is measured, is constructed with reference to the reciprocal relations between electricity and magnetism, the units are usually termed *electromagnetic* to distinguish them from *electrostatic* units, which represent electric force alone, and from *magnetic* units, which represent magnetic force alone.

Instruments for electric measurement are constructed either on the principles of electric attraction and repulsion, on the relations between electricity and magnetism, on the heat developed by the electric current, or on the amount of metal deposited or gas generated by electrolysis.

Electrometers.—The instruments by which electrostatic force is measured are known as *electrometers*, and measure either the absolute force by which one electrified body attracts another by direct movement, as in the attracted-disk electrometer; or the relative force by which one repels another by a rotary movement, as in the torsion balance; or the combined relative effects of attraction and repulsion by rotary movement, as in the quadrant electrometer. As all these electrostatic instruments and methods of measurement are fully described in the author's "Elements of Static Electricity," further reference to them here is unnecessary.

Galvanometers.—Instruments for electric measurement constructed on the principle that the magnetic needle tends to assume a position at right angles to that of the electric current were formerly known exclusively as *galvanometers*, a term still applied to the older instruments of this class, while certain improved instruments recently constructed on this principle are known as voltmeters and ammeters, the former used to measure electromotive force, and the latter current strength.

Instruments indicating the presence and direction of electric currents, as the galvanoscope, Schweigger multiplier, and astatic needle, have already been described in Chapter IV, but none of these measure current strength, though roughly indicating its amount, while the galvanometer, constructed on the same principles, is a much more accurate instrument. Its general construction consists of a magnetized needle, poised so as to have a free horizontal rotary movement, and inclosed within a coil of insulated copper wire through which the electric current can flow; the strength of the current being measured by the needle's deflection as shown on a graduated circle of 360°.

Galvanometers are adapted only to the measurement of direct currents, and are but slightly affected by alternating currents.

In every galvanometer except the astatic, the needle and the vertical plane of the inclosing coil are set in the plane of the magnetic meridian, so that the deflecting force of the current acts at right angles to the horizontal component of the earth's magnetism, the former tending to rotate the needle into a position at right angles to the direction of the latter. Hence it is evident that the amount of the deflection never can exceed 90°, since at this angle the position of the needle is normal to the direction of the current, and the force represented by the angle at a maximum.

But the deflecting force does not vary in the same ratio as the angle of deflection, since the needle receives the full effect of this force only when in the vertical plane of the coil, which in this case coincides with that of the magnetic meridian, while in every other position only a portion of this force acts on it, and the strength of this effective portion varies inversely as the angle of deflection.

This matter will be better understood, especially by

those not familiar with the measurement of angles, by reference to Fig. 45.

Let the line NS represent the needle in the plane of the magnetic meridian, poised at its center C, so that it can be rotated by the deflecting force into the position WE, or any intermediate position; the force acting on its north pole, N, tending to rotate it toward E, and that acting on its south pole, S, tending to rotate it toward W. When the needle has thus been turned from the position NS, the deflecting force acts on it obliquely, its

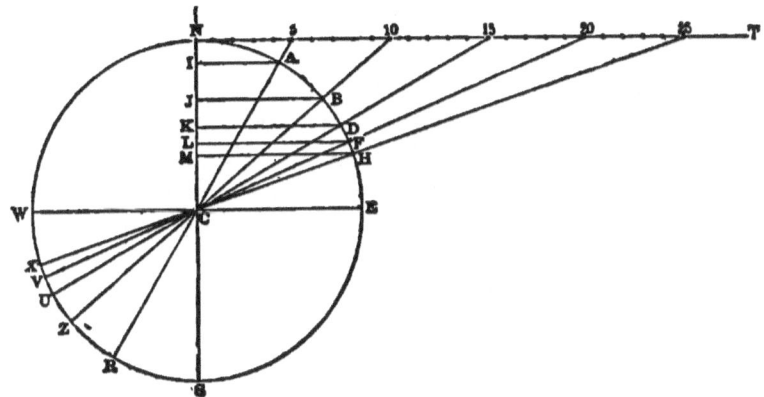

FIG. 45.

effective component on the north pole, when in the position AR, being represented by the line IC, while the remainder acts along the needle's length, and is not represented by the angle of deflection; a similar result being true of the deflective force on the south pole; hence the effective part of this force in the position AR is to that in the position NS, as IC to NC; and so when the north pole has been deflected to B, D, F, or H, the effective part, as compared with that represented by

NC, is represented respectively by the lines JC, KC, LC, and MC.

But the effective part represented by each of these lines belongs to a current of increased strength, otherwise it could not produce the increased deflection, and hence, though representing a constantly decreasing increment of the total force, its actual strength is increasing directly as the angle of deflection; so that the effective part, represented by the short line MC, is as much stronger than the entire deflective force represented by NC as the angle NCH is greater than zero.

Measurement of Angles.—Angles are measured by certain functions known as sines, cosines, and tangents. Take any angle, as NCA, and with any part, NC, of one of the inclosing lines, as radius, and the point C, where the lines meet, as center, describe a circle; and from the point A, where the other inclosing line meets or intersects the circumference, draw a line, AI, perpendicular to NC; the ratio between the length of this line and radius is the *sine* of the angle. And the length of radius being taken as the unit, the sine is represented by the length of this perpendicular, which therefore is the measure of the angle. Hence each of the lines, BJ, DK, FL, HM, perpendicular to NC, is, like AI, the measure of the angle which it subtends. The length of this perpendicular may vary from radius to zero, but evidently can never exceed radius.

The *cosine* is the ratio between the length of radius and that part of it included between the center and the point where the perpendicular representing the sine meets it. Hence, radius being unity, CI represents the cosine of the angle NCA, and may also be taken as its measure; the value of the cosine varying inversely as that of the sine. In like manner CJ, CK, CL, and CM

represent the cosines of the other angles mentioned, and hence measure them.

The *tangent* is a straight line which touches the circumference of a circle, or arc, at any point, but which, if produced, does not cut it, as NT; and hence it forms a right angle with radius at the point of contact. For any angle less than a right angle, it is included between the line coinciding with radius at the point of contact and a straight line drawn from the center and produced to meet it, and hence it subtends the angle formed by these lines. Thus N_5 is the tangent of the angle NC_5, and each of the lines, N_{10}, N_{15}, N_{20}, and N_{25}, the tangent of the angle which it subtends.

The length of the tangent varies from zero to infinity; the tangent of a right angle being infinite, since it is perpendicular to one of the inclosing lines and parallel to the other, and hence can never meet the latter.

Angular Measurement of Deflective Force.—It has been shown in Chapter IV that the horizontal force of the earth's magnetism, by which the needle is deflected, must vary as the tangent of the angle of deflection; but in the galvanometer this force is represented by that of the current flowing in the coil, hence the same rule applies; so that if the tangent be laid off into equal spaces, as in the figure, and lines from the dividing points be drawn to the center, those spaces must represent equal increments of current strength, though the increments of the circumference included between these lines and also the cosine, which represents the effective component of the deflective force, constantly decrease as the angle of deflection increases. Hence the total deflective force, representing the current's strength, does not vary as the arc through which the needle rotates, but as the tangent of the including angle.

For instance, the ratio of strength between a current

producing a deflection of 10° and one producing a deflection of 20° is not that of 10° to 20°, but of tan 10° to tan 20°; for it requires a current of much more than double the strength to double the arc, since, as already shown, only that portion of the total force represented by the cosine is effective in producing the deflection; but the tangent of 20° is much more than twice the length of the tangent of 10°, and represents the total increment of force, effective and non-effective, while the cosine represents only the effective portion. Now since it has been shown that the strength of the effective increment varies as the angle, it is correctly represented by the angle's sine.

Hence the EFFECTIVE *deflective force varies as the* COSINE, *its* STRENGTH *as the* SINE, *and the* TOTAL STRENGTH OF CURRENT *as the* TANGENT *of the angle of deflection*.

Calibration of Galvanometer.—A galvanometer may be *calibrated* by ascertaining from comparison with a similar standard instrument, or otherwise, the different degrees of current strength represented by different degrees of deflection; and these results being tabulated are a correct guide for the use of the instrument so long as the magnetic strength of the needle remains unimpaired, and the functions of other parts affecting the deflection remain constant.

Melloni used the differential deflections of opposite electric currents produced by heat as a means of calibration; and the term seems, perhaps for this reason, to have been derived from thermometric calibration, to which it is analogous.

All instruments for electric measurement require jewelled bearings for the rotating parts, to reduce the friction to the minimum.

Sine Galvanometer.—The deflective force may be measured either by the sine or the tangent, according to the construction of the galvanometer. Where great sensitiveness is required the *sine* galvanometer is preferred. Its essential features are a long needle and an inclosing coil of only sufficient diameter to permit the needle's free oscillation, and which can be rotated horizontally. Its construction will be understood from Fig. 46.

The needle is mounted at the centre of a vertical coil,

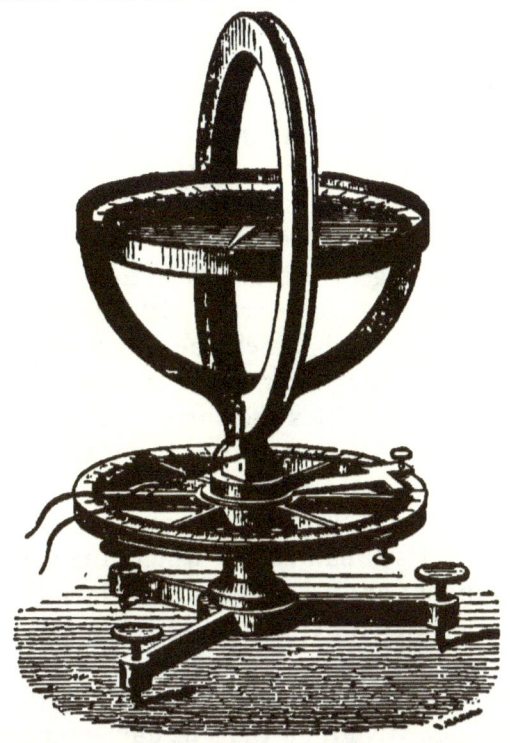

FIG. 46.

composed of a number of convolutions of insulated copper wire wound on a circular grooved brass frame, and underneath the coil is mounted a circle, graduated to correspond to a similar one in proximity to the needle above; the centre of each being in a vertical line with the centre of the needle. The upper circle is attached to the frame of the coil, so that both can be moved horizontally by an index lever, shown below, through any number of degrees indicated on the lower circle by the index.

The instrument being set with the plane of the coil in the magnetic meridian, parallel to the needle, which points to zero, and a deflection being produced by the passage of the current to be measured, the needle rotates out of the plane of the coil to a position where the magnetic field is weaker; the coil is then turned in the same direction as the needle, its approach producing further deflection, till its plane again coincides with the needle, which again points to zero. In this position the deflective force of the current is evidently just equal to the opposing horizontal force of the earth's magnetism, which would bring the needle back to its original position if the deflective force were withdrawn. The number of degrees through which the coil has been turned being noted on the lower scale, the sine of the corresponding angle indicates the current strength, to which it bears a certain definitely varying ratio, as has been already shown.

This mode of measurement is approximately accurate for angles of less than 20°, in which the values of the sine and tangent are nearly equal, but is not reliable for larger angles, the difference in those values being too great, as has been shown.

Tangent Galvanometer.—The *tangent* galvanometer, though less sensitive than the sine galvanometer, is simpler in construction and more accurate for large deflections by strong currents, and hence is generally preferred. Its essential features are a short needle, and a coil of relatively large diameter, varying from ten inches to one meter. The needle is usually about three quarters of an inch in length, diamond-shaped, with an aluminium pointer of convenient practical length attached at right angles to its polar diameter. The large diameter and circular form of the coil are to create a field of approximately uniform strength within the small central area in which the needle rotates; the inner lines of force from the current converging to the centre; and the needle is made short so as to be confined as closely as possible to this small central space, where the field is most uniform.

Fig. 47 shows an instrument of this class; its coil consisting of a single turn of copper wire, having prac-

FIG. 47.

tically no resistance, and not requiring a supporting frame. The coil terminals are connected with two tubes shown at the base, one inside the other, insulated from each other and furnished with binding-screws. The needle-case is from four to five inches in diameter.

The tangent values are sometimes laid off on the scale of the instrument in the manner shown in Fig. 45;

128   *DYNAMIC ELECTRICITY AND MAGNETISM.*

but as it is difficult to do this with requisite accuracy, the scale of degrees is usually preferred, the values of the tangents corresponding to the deflections being easily ascertained from a table.

The Helmholtz-Gaugain tangent galvanometer is illustrated by Fig. 48. The needle is placed at the centre of a straight line connecting the centres of two separate coils of equal size, set parallel to each other at a distance apart equal to their radius. This arrange-

FIG. 48.

ment insures much greater uniformity of field than can be obtained from a single coil, and still greater uniformity could be obtained by the addition of a third coil, midway between the two, and of such diameter that each of the three should be equally distant from the centre of the needle. The two-coil method was proposed by Helmholtz, while Gaugain proposed placing the needle in the same relative position on one side of a single coil.

The instrument here shown has two sets of coils, marked *A* and *B*, four in all, connected with binding-screws at the base of each circular support, as shown, one coil of each set on each support. The *A* set has very low resistance, only a small fraction of an ohm to each coil; each being composed of about four turns of No. 12 copper wire. The *B* set has very high resistance, 10 to 12 ohms to each coil, each composed of No. 26–30 wire. The needle is suspended by an untwisted silk fibre inclosed in a vertical tube, and adjusted by the screws shown at top, so that it rotates without friction and against only slight torsion.

Astatic Galvanometer.—A very sensitive galvanometer, originally invented by Nobili, may be constructed with the astatic needle described and illustrated on page 73; which, being approximately independent of the earth's magnetism, is deflected by a very slight current. Fig. 49 shows the construction.

Two short needles with poles reversed are attached to a common support, which also carries a light pointer of convenient length. The coil is flat and usually of sufficient horizontal diameter to inclose the lower needle entirely in any position; while the upper needle rotates over its upper surface, and the pointer over a dial-plate with scale above.

FIG. 49.

The needle is suspended at the centre of the coil from a vertical support by a single fibre of silk, or by two parallel fibres hung near each other; the latter method being known as *bifilar suspension*, its object being to bring the needle to rest in a fixed position more perfectly than can be done by the torsion of a single fibre; the needle being raised slightly when, by its deflection, the two threads are twisted out of parallelism, and its weight tending to bring them back to the parallel position. The suspension is adjusted by the thumb-screw shown above; the needle being set parallel to the vertical plane of the coil; and as it is impossible to make a needle perfectly astatic, both should also be parallel to the plane of the magnetic meridian. A glass shade affords protection from air-currents.

The readings, for the reasons already given, are only approximately accurate, and for deflections greater than 20° unreliable; but the instrument can be calibrated for larger deflections.

Thomson's Reflecting Galvanometer.—This instrument, invented by Sir William Thomson for telegraphing through long submarine cables, is exceedingly sensitive. Its construction is shown by Fig. 50; its principle being practically that of a tangent galvanometer with a long pointer and tangent scale.

At the center of a line connecting a pair of small coils of equal size and resistance, is suspended, by a silk fibre, a diminutive concave mirror, of about 1 centim. diameter, with a little needle, made usually of a piece of watch-spring, attached to its back; the weight of both not exceeding one or two grains. A small circular opening in the case, directly opposite the mirror, widening outward, admits the light from a lamp connected with the graduated scale shown in Fig. 51.

This scale is placed in front of the galvanometer, at a distance of about 36 inches, and the lamp is placed in the box at the right which excludes the direct rays. The light is transmitted through a tube terminating in a small circular opening, from which a beam falls on the small mirror shown just below the centre of the scale, and is reflected to the galvanometer mirror, and thence back to the scale; the mirror adjustments being such that a small spot of light, concentrated by a lens in the galvanometer, shown in the cut, is reflected on

FIG. 50.

zero of the scale, when no current is passing, and moved to the right or left, according to the direction of the current, to a distance corresponding to the current strength. The shadow of a fine wire, stretched in front of the galvanometer mirror, indicates the exact centre of the spot of light, which is adjusted to zero by a curved magnet, attached above to a vertical rod, with its poles in opposition to those of the needle, and which can be moved to any required position vertical or horizontal.

The pointer being the ray of light, 36 inches long, the slightest deflection is prominently indicated on the scale; a current produced by dipping the points of a

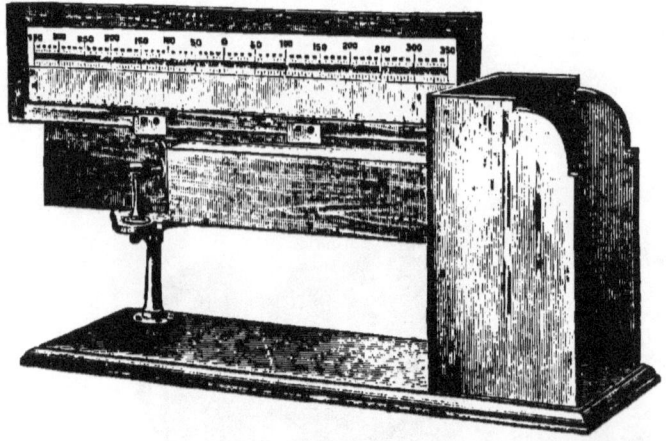

FIG. 51.

brass pin and a sewing-needle into a drop of salt water, moving the spot of light half the length of the scale.

The coils can be removed and coils of any required resistance up to 5000 ohms substituted; and as these and similar delicate measuring instruments are liable to injury from powerful currents, which also produce deflections too great for accurate measurement, *shunts* of fine wire are provided, separate from the instrument, by which fractions of the current, of measurable strength, are transmitted ; and the respective resistances of coil and shunt being known, the entire current strength can be ascertained.

The requisite light can be furnished by an ordinary kerosene lamp, but that of an electric lamp or a lime burner, when obtainable, is far superior.

Fig. 52 shows another style of the same instrument

FIG. 52.

with four coils in two sets, upper and lower, having any required resistance up to 8000 ohms.

Differential Galvanometer.—This instrument is constructed with two coils of equal size and resistance, between which the needle is mounted at the central point, and through which currents may be transmitted simultaneously in opposite directions and their relative strength compared : if equal, there is no deflection ; but if unequal, the relative difference in strength is shown by the amount of the deflection.

Ballistic Galvanometer.—A ballistic galvanometer is one constructed with a needle weighted by inclosing it in lead or otherwise, so that the impulse given it by a transient current of too short duration to be measured in the ordinary way may be developed slowly by the needle's momentum, so that the amount of deflection can be more easily observed.

When used to measure current quantity, as indicated by current strength in the discharge of a condenser, the sine of half the angle of deflection produced by the first swing of the needle is taken as proportional to the quantity of the transient current thus produced.

Common Galvanometers.—Galvanometers of various styles and sizes are constructed for ordinary practical use, usually with flat coils of various degrees of resistance. Such instruments are often better adapted to measurements where only approximate accuracy is required than those of finer construction, but are not suitable for strict scientific work.

Voltmeters and Ammeters—Galvanometers measure only current strength, usually in degrees of an arc, but it has become important in the progress of electric development to measure also electromotive force, and to express the measurements of both E. M. F. and current strength in volts and amperes, either directly or in terms easily reducible to those units : for this purpose voltmeters and ammeters are constructed.

The difference between these two instruments consists chiefly in the respective resistance of each, and its relative position in use ; the voltmeter having high resistance and being placed in a derived circuit between the points whose difference of potential is to be measured, while the ammeter has low resistance and is placed directly in the main circuit at any point where current strength is to be measured.

It will be noticed that an unmagnetized, soft-iron needle, or armature, is an important feature of many of these instruments.

The Weston Voltmeter.—This instrument, shown by Fig. 53, incloses within its case a powerful steel horse-

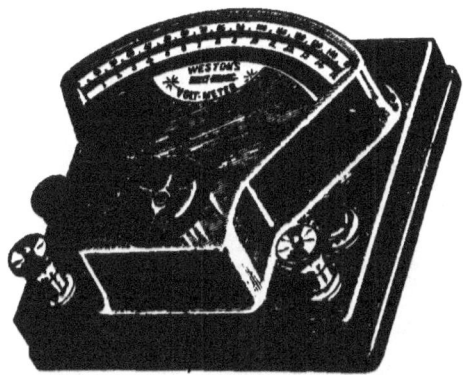

FIG. 53.

shoe magnet, the poles of which project into the narrow space in front and are attached to two soft iron pole-pieces, as shown in Fig. 54. These inclose a circular space, within which is mounted a soft-iron armature core, maintained in a fixed central position by attachment to a brass yoke which connects the pole-pieces; part of this yoke, with its right-hand connection and a central projection for attachment of the core, being shown.

A light copper frame, $\frac{3}{8}$ of an inch wide, and wound with a coil of fine, insulated copper wire, surrounds the core, and has a limited rotary motion, on jewelled bearings, in the narrow space between the core and pole-pieces, which is just wide enough to allow rotation without contact.

The terminals of the coil are connected above and below with two flat springs, oppositely coiled, and so attached to the copper frame and adjoining parts as to maintain the coil in a fixed position, when the springs a‧e not under tension, and bring a light aluminium

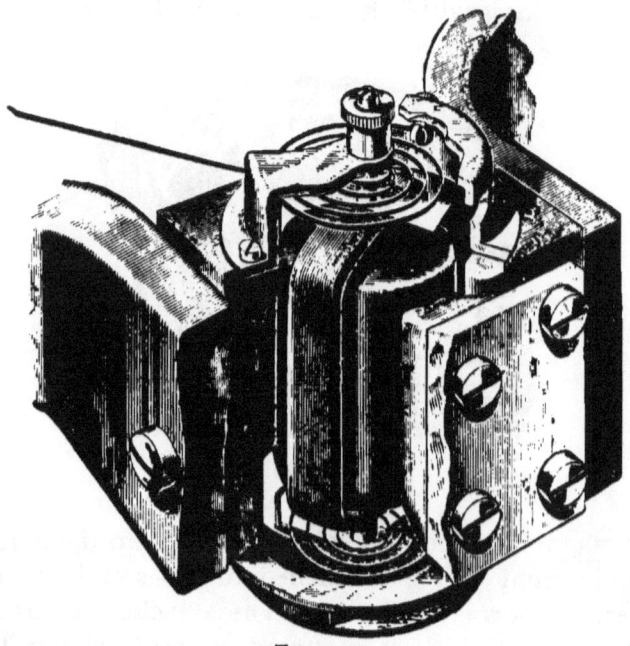

FIG. 54.

pointer, attached to the frame, to zero of the scale on the left.

These springs are made of a special, non-magnetic alloy, and are placed in opposition to neutralize the effects of expansion and contraction under variations of temperature.

A resistance coil, mounted within the case, makes electric connection, by one of its terminals, with one of

ELECTRIC MEASUREMENT. 137

the springs, while the other terminal is connected with the front binding-post on the left. Another connection with the rear binding-post on the same side taps this coil at a point nearer the spring, so as to include a much lower resistance. The other spring is connected with the binding-post on the right, back of which is a contact key and a calibrating coil. This part of the circuit can be closed permanently, after calibration, by depressing the key and giving it a quarter-turn.

When connections with an electric source are made by the right binding-post and either of the two on the left, the current enters and leaves the copper coil through the springs, its direction and the winding being such as to produce deflection from left to right; the coil tending to rotate into a position at right angles to the lines of magnetic force, in opposition to the tension of the springs. And the instrument being calibrated in accordance with the resistance of its coils, the deflection of the pointer will indicate the difference of potential in volts; since with a given resistance the E. M. F., or potential difference, varies directly as the current strength.

The entire resistance is to that of the sectional part in the ratio of 20 to 1; the divisions of the scale being in volts for the outer reading, corresponding to the high resistance, and the same in twentieths of a volt for the inner reading, corresponding to the low resistance, as shown. Hence the E. M. F. which will produce a deflection of one division, when connection is made with the front binding-post on the left, will produce a deflection of twenty divisions when connection is made with the rear binding-post.

The high-resistance circuit is used for apparatus generating strong currents, as dynamos, and the low-resistance circuit for apparatus generating weaker currents, as primary batteries, on account of its greater sensitiveness: and as a dynamo current would be likely to injure or destroy the copper coil, if admitted through the low resistance, the rear post is protected from accidental contacts by an outer covering of hard-rubber. In some of the instruments all the posts are similarly protected; the rubber also preserving the contacts from oxidation. The scale readings also vary in different instruments.

The deflection of a current-bearing coil in a magnetic field of special strength gives this instrument great superiority over instruments depending on the deflection of a steel or soft-iron needle; the magnetic action being stronger, and its relation to the current more direct. The constancy of the instrument is dependent solely on the constancy of the magnet, the springs, and the internal resistance.

The Weston Ammeter.—The construction of the Weston ammeter is similar to that of the voltmeter, but simpler; the chief differences being that the copper coil is of coarser wire, having much lower resistance, and the resistance coil is not required: hence there are only two binding-posts and a single circuit, directly through the copper coil and springs.

The scales for different instruments range from 5 amperes, with divisions of $\frac{1}{20}$ of an ampere, to 100 amperes, with divisions of 1 ampere, according to the relative resistance of the coils.

The Weston Milliammeter.—This instrument has the same construction as the ammeter but lower resistance. Instruments of two different resistances, with scales of corresponding difference, are constructed; one of 300 milliamperes, with scale divisions of 2 milliamperes each; and the other of 600 milliamperes, with scale divisions of 4 milliamperes each.

A milliampere being $\frac{1}{1000}$ of an ampere, it is evident that these instruments are capable of measuring very low currents, especially as the scale divisions are readable to fifths; so that the smaller instrument can indicate a current of $\frac{1}{5}$ of 2 milliamperes, $= \frac{1}{2500}$ of an ampere.

The Wirt Voltmeter.—This instrument, illustrated by Fig. 55, is constructed on the principle of ascertaining the E. M. F. to be measured by comparison with a known

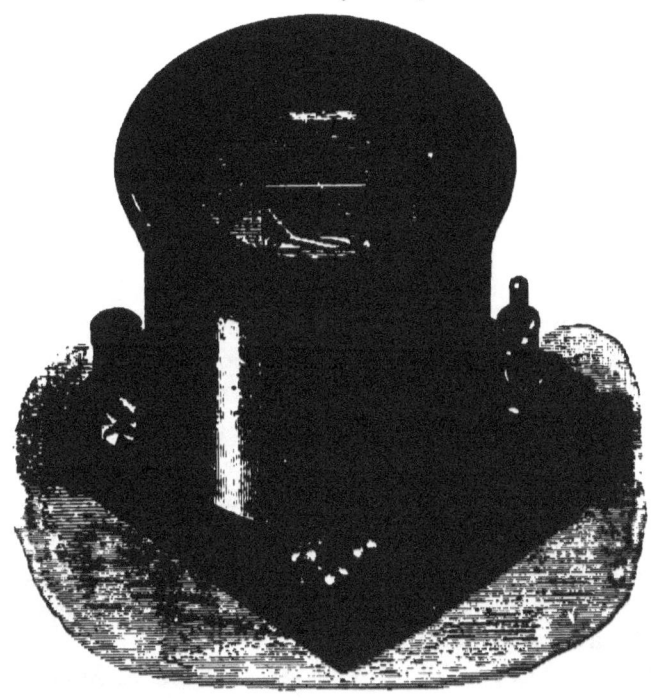

FIG. 55.

E. M. F.; each being proportional to a resistance having similar conditions through which the measurement is made.

The case incloses two Clark cells, each having a constant E. M. F. of 1.43 volts, the connections being so arranged that either can be employed alone, or the two joined in series so as to obtain an E. M. F. of 2.86 volts.

Under the glass cover is shown a small galvanometer, with magnetic needle, light aluminium pointer, and terminal wires connected with the coil; also a small scale, not shown, under the pointer, having a limited range, in opposite directions, from o at the centre.

Extending round the case inside is a coil of german-silver wire, having a resistance of about 2500 ohms, one terminal of which is attached to one of the binding-posts shown on the right, marked $+$, while a sliding contact, which can be moved to any required point on this coil, is connected with the other binding-post, marked $-$; and this contact is attached to the rim of the hard-rubber cap, shown above, which can be rotated on the interior part of the cap, on which is shown a scale graduated in volts, from $1\frac{1}{2}$ to 120. By rotating this rim, a short index, attached to it, is moved to any required point on the scale, the sliding contact being moved simultaneously, so as to include any resistance required between the terminals of the binding-posts.

The galvanometer circuit also includes a certain portion of this coil, having a known resistance calibrated with reference to the known E. M. F. of the battery cells, which are also included in this part of the circuit. A contact key, shown on the left, closes this circuit through the galvanometer, producing deflection of the needle and attached pointer.

If connection with a generator whose E. M. F. is to be measured be made through the binding-posts, so that the current shall oppose the meter's battery current, the needle will be deflected, when the contact key is closed, so long as the generator current is stronger or weaker than that of the battery.

Let the instrument be so placed that the earth's magnetism shall bring the galvanometer pointer to o on the small scale; and let the rim be turned so as to bring the attached index near the probable E. M. F. on the large scale; then, deflection being produced by closing the contact key, let the rim be turned so as to include sufficient resistance to equalize the opposing currents and bring the galvanometer pointer back to o; the index will then show the E. M. F. of the generator in volts on the large scale. For, since with a given current, E. M. F. varies directly as resistance, if the E. M. F. of the battery be represented by E and that of the generator by E', the resistance of the battery circuit by R and that of the generator circuit by R', then $R : R' :: E : E'$. That is, the resistance of the battery circuit is to the resistance of the generator circuit as the E. M. F. of the battery is to the E. M. F. of the generator, and the calibration gives this E. M. F. in volts.

A switch is shown in front by which connection can be made with either of two separate circuits, the right-hand contact, marked $\frac{1}{10}$ to indicate the relative measurement of E. M. F., connecting with one having ten times the resistance of that connected with the left-hand contact. At the opposite corner, in the rear, three battery connections are arranged, the right and left ones, marked A and B, being each through a separate cell, and the central one, marked 2, through the two cells in series; a plug closing whichever connection is to be used. When the switch is on contact 1, as shown, and the plug in A or B, the scale readings require no correction, and should be the same with the plug in either hole, each cell being a check on the accuracy of the other. But when the plug is in hole 2, the cells being in series, the reading must be multiplied by 2, since the battery E. M. F. is doubled; for $R : R' :: 2E : 2E'$.

But when a generator of low E. M. F. is to be tested, the switch is connected with the contact marked $\frac{1}{10}$, which includes, in the battery circuit, a resistance of ten times that included by contact 1; hence, since the battery current with this resistance is only $\frac{1}{10}$ of what it was with the former resistance, $\frac{1}{10}$ the E. M. F. will develop an opposing current of equal strength, giving the same reading, which must be divided by 10 to give the correct E. M. F.; for $10R : R' :: 10E : E'$.

Each cell is $1\frac{3}{4}$ inches high and $\frac{5}{8}$ of an inch in diameter, constructed with an inverted glass cup, inclosed in a brass case and hermetically sealed with soft rubber melted into the bottom.

The electrodes are zinc and mercury, and the fluid zinc sulphate and mercuric bisulphate, formed into a paste in which the electrodes are inclosed; connection with the mercury being made by an insulated platinum strip which represents the positive pole.

This cell is selected on account of the remarkable constancy of its E. M. F., and the instrument is calibrated for a cell temperature of 21° C., requiring a correction in the reading of .000367 per degree of variation above or below 21° C., which must be made by subtraction for the higher temperature, and by addition for the lower.

The cells are easily removed and replaced, when necessary, without disturbing the connections; and being small, hermetically sealed, and amply protected, do not interfere in the least with the handling of the instrument, and can be cheaply replaced when exhausted.

Ayrton and Perry's Spring Voltmeters and Ammeters.—The unreliability of electric measuring instruments constructed with permanent magnets, liable to magnetic loss, or to variation of magnetism from the influence of powerful currents, and consequently requiring frequent recalibration, has led to improved methods of construction, of which the spring voltmeters and ammeters of Ayrton and Perry are a result. Fig. 56 represents the ammeter, the voltmeter being of similar construction;

the principle being simply the torsion of a spring by electromagnetic attraction.

The current passes through a long, narrow vertical coil, of high resistance in the voltmeter and low resistance in the ammeter, within which is suspended a light soft-iron tube, which incloses a long spiral spring of phosphor-bronze ribbon. This spring supports the tube, being attached at bottom to a brass cap in which the tube terminates, and above to a milled head which rests on the glass cover and is connected with the spring by a

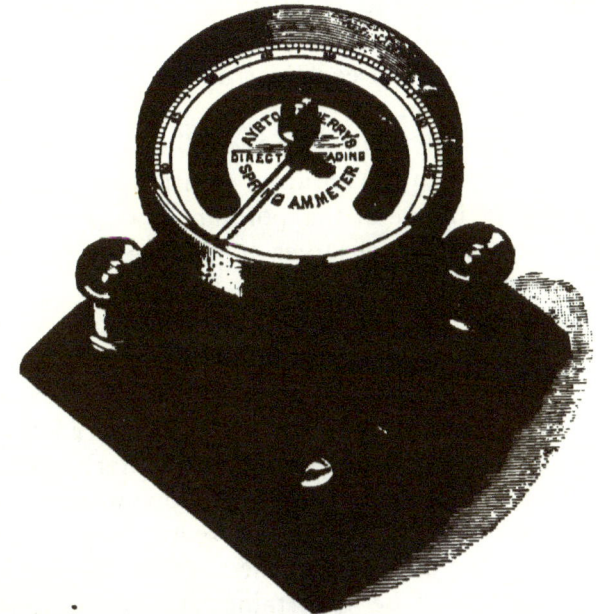

FIG. 56.

vertical pin which passes through the glass; a similar pin projects downward from the bottom of the brass cap and passes through a hole in a support below, in which it has a free vertical movement; so that the two pins hold the spring and tube in a vertical position; and the tube being shorter than the coil, its centre on a vertical line is above that of the coil. To the top of the tube it attached a light pointer which rotates over a scale graduated either in volts or amperes according to the design of the instrument.

When no current is passing the pointer indicates zero on the left of the scale, but when the current passes, the tube is pulled down by magnetic attraction, in opposition to the torsion of the spring, to a distance proportional to the current's strength; giving it a rotary motion by which the pointer is deflected, which indicates by direct readings the E. M. F. in the voltmeter, and the current strength in the ammeter, according to the respective resistance of each instrument, and its position in the electric circuit.

The tube can be turned by the milled head so as to bring the pointer to the required position in calibrating; and a reflected image of the pointer, in a mirror placed under it, enables the observer to determine accurately its position on the scale.

A little magnetic needle, shown at the front corner of the base, indicates the direction of the current; but as such a needle is liable to have its poles reversed by powerful currents, a bar magnet is preferred for this purpose. Since the deflection of the pointer depends on the magnetic attraction of the tube downwards, it must evidently be always in the same direction, and hence independent of the direction of the current; so that while this direction may be ascertained as above, it is not essential to the use of the instrument that it should be known.

A light movable auxiliary coil surrounds the main coil and is connected with it in parallel; this can be moved up or down in calibrating till a position is reached in which its inductive influence on the main coil is best adapted to the construction, where it is made stationary.

The case is ventilated, as shown, to prevent the accumulation of heat generated by the current, which would expand the spring and produce inaccuracy. The usual binding-posts connected with the terminals of the coil are shown at the right and left, the left post being marked A to distinguish them in use.

The voltmeters are usually constructed to measure E. M. F. ranging from 15 volts to 1000; the ammeters, to measure current strength ranging from $\frac{1}{10}$ of an ampere to 600 amperes.

Gravity Ammeters.—While springs have greater constancy than permanent magnets in the construction of electric measuring instruments, their constancy is liable to vary, or be impaired, from well-known causes, as heating, age, and use, imperfect material, or oxidation; but the force of gravity, being always known and constant, may be utilized in such construction to produce instruments of great constancy. On this principle the United States Electric Lighting Company constructed the ammeter shown in Fig. 57.

Two pairs of electromagnets, wound with coils of low resistance, and having laminated soft-iron cores, are placed as shown; each pair having its coils wound on the same core, producing consequent poles, but magnetically insulated from the other pair.

At the centre, between these magnets, is mounted a soft-iron armature, lightly poised on a horizontal axis, the end of which is seen through the circular opening, and having a vertical rotary movement parallel to the magnets' plane. This armature is about 2 inches long, $1\frac{1}{2}$ inches wide at each end, $\frac{3}{8}$ of an inch at the centre, and $\frac{1}{4}$ of an inch thick; its sides concave, and its ends convex and slotted to correct the effects of residual magnetism. A pointer, attached to its axis, indicates the readings on a scale above, as shown.

When no current is passing, the armature is maintained in a fixed position by one or more little weights attached to its lower left-hand corner, its longer axis

FIG. 57.

being on a diagonal line between the lower left and upper right-hand corner of the instrument, and the pointer at zero on the left of the scale. But when the current passes through the coils in either direction, the armature rotates in obedience to the electromagnetic force, its longer axis tending to assume a horizontal

ELECTRIC MEASUREMENT.   147

position, and the pointer is deflected from left to right in proportion to the current strength, which is indicated by direct reading in amperes.

By the removal or addition of one or more of the little weights, the sensitiveness of the instrument may be varied in calibrating, as required for different ranges of current strength. The terminals of the coils are shown at the base, and holes for ventilation at the top of the case.

Instruments constructed on this principle have not been employed to any great extent as voltmeters, not being sufficiently sensitive for the light currents required.

Since the weight, as it rises, recedes from the vertical line which passes through its axis of rotation, the force opposing rotation increases in the direct ratio of the increase of leverage thus produced. Hence, as equal divisions of the scale would represent unequal increments of current strength, they should be made in the inverse ratio of this increase of leverage.

But as it is difficult to mark off such short spaces with the requisite accuracy, a gravity ammeter has been constructed by the Western Electric Company, with a vertical electromagnet having a pole-piece so curved that the rotating armature, as it rises, constantly approaches it, the magnetic attraction increasing in the same ratio as the leverage, so that equal divisions of the scale represent equal increments of current strength.

The Cardew Voltmeter.—The instruments thus far described are designed to be used with direct currents, and are liable to errors arising from self-induction in addition to those from the other causes mentioned. But since, according to a well-known law, the heat developed in an electric conductor is in direct proportion to the square of the strength of the current passing through it, instruments can evidently be constructed on

148 DYNAMIC ELECTRICITY AND MAGNETISM.

this principle which will measure either current strength or difference of potential, produced either by direct or alternating currents, and are not liable to variation from any of the causes mentioned. Among these the voltmeter, patented by Cardew in 1886, has a prominent place. Its operation depends on the expansion of metal produced by the electric development of heat.

Fig. 58 gives a front view of this instrument and Fig. 59 a rear view, showing its internal construction. A fine

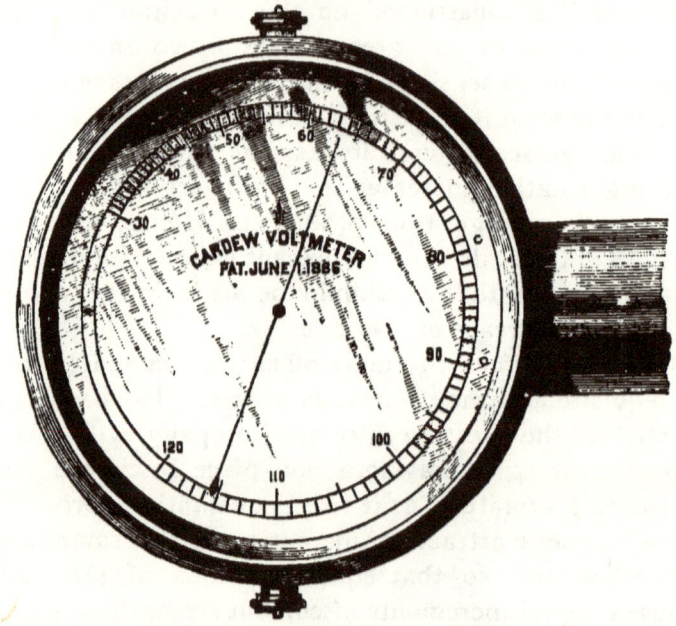

FIG. 58.

platinum wire, 8 feet long, is stretched in four lengths in a horizontal tube, by attachment to a metal frame and pulleys, as shown at a, a, t, t in Fig. 59. This tube is made of very thin metal, one third of its length being iron and two thirds brass, to maintain constancy of length between the points of attachment of the wires by such a mode of connection as to produce compensation by the unequal expansion of the two metals; and the horizontal position is given it to maintain constancy of temperature, and prevent the unequal expansion, from

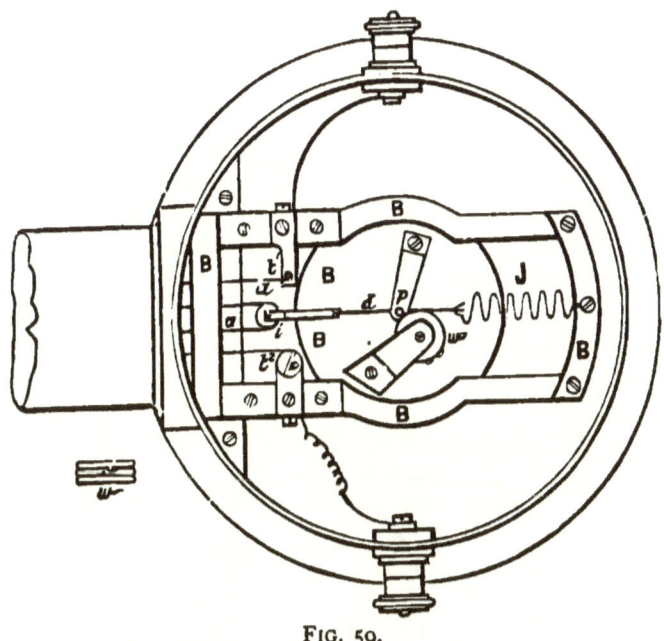

FIG. 59.

convection of the air to which the tube and wire would be liable in a vertical position.

The wire has a resistance of about 240 ohms, and attains a maximum temperature of about 200° C.; and its expansion varying in a certain definite ratio dependent on the difference of temperature caused by the passage of the electric current, which, as stated, varies as the square of the current's strength, produces a variation in length proportional to the E. M. F. by which the current is generated. This produces a rotation in the pulley w, to the axis of which the pointer shown in Fig. 58 is attached, which moves in the same direction as

150 *DYNAMIC ELECTRICITY AND MAGNETISM.*

watch-hands when the E. M. F. increases, and in the opposite direction when it decreases.

This instrument should be calibrated for the average temperature of the room in which it is to be used.

The Edison Current-Meter.—Instruments for measuring the amount of electric current used by a consumer of light or power are constructed on various principles. Among these is the *Edison current-meter*, in which a small percentage of the current is passed through two cells containing amalgamated zinc plates immersed in a solution of zinc sulphate. Zinc is thus deposited on the plates, which are removed and weighed at stated times, and the consumption of current being in proportion to the amount of deposition, according to the principle discovered by Faraday, is estimated accordingly.

FIG. 60.

The Forbes Coulomb-Meter.—Meters like the Edison

cannot be used for the measurement of alternating currents; but one has been invented by Forbes, operated by the heat developed by the current, which can measure either direct or alternating currents. Its construction is shown in Fig. 60. The current passes through a flat coil of iron wire, above which is mounted, on a paper cone having a jeweled bearing at its apex, a mica disk, with mica vanes attached. The heat developed by the current produces an ascending current of air which rotates the disk, operating a light train of clock-work which moves indexes over two dials, registering the current consumption in coulombs; units being registered on one dial and tenths on the other. A glass shade protects the apparatus from external air-currents.

Voltameters.—Instruments like the Edison current-meter are more generally known as *voltameters*, a name given them by Faraday, who first proposed this method of electric measurement. They may be constructed with any substance practically susceptible of electrolysis, in accordance with Faraday's law that the amount of an element liberated by electrolysis in a given time is proportional to the strength of the current employed. Salts of copper and of silver are both employed for this purpose, also acidulated water.

The Water Voltameter.—This is simply a common decomposing instrument in which the liberated elements, oxygen and hydrogen, are collected in the same receiver, which is graduated in cubic centimeters or any other convenient standard. The amount of each gas produced at a standard temperature and pressure, by a coulomb of electricity, being known, the entire number of coulombs consumed in a given time can easily be ascertained. This amount, at temp. 0° C. and press. 760 millims., is found to be 0.0579 cubic centims. of **oxygen**

and 0.1157 of hydrogen, making 0.1736 c.c. of both, per coulomb of electricity.

The use of such an instrument is confined to the laboratory, as the wasteful consumption of current, the resistance due to polarization, and the loss from recombination of the gases, or escape of the hydrogen, renders it unsuitable for practical measurement.

The Weber-Edelmann Electrodynamometer.—This instrument, invented by Weber and improved by Edelmann, is constructed on the principle of the deflection of a coil, in opposition to the torsion of a wire, by the joint product of E. M. F. and current strength.

Fig. 61 shows the construction. Two coarse wire coils of low resistance are mounted parallel to each other on a stand, on three transverse brass rods, supported by a vertical brass ring, at the centre of which is suspended a small, fine wire coil of high resistance; its plane, when at rest, being at right angles to the planes of the larger coils. A small plane mirror is attached to the centre of the small coil, to which a ray of light from a lamp is admitted through an aperture in the little screen shown in front of it.

The suspension of the small coil is by means of a wire connected with its terminals and inclosed in the vertical brass tube shown. This wire is attached to the projecting rods seen at the top of the upper section of the tube and the bottom of the lower section; the set-screws and nuts shown being used to give proper adjustment to the coil and tension to the wire; the terminal rods passing through movable disks for this purpose.

The current from the generator enters by one of its circuit terminals, attached to a binding-screw at the bottom of the lower section of the tube, passes up through the inclosed wire and traverses the small coil, goes thence through the upper section of the wire and

ELECTRIC MEASUREMENT. 153

returns by the upper section of the tube to the ring, passes through one of the rods to a terminal of one of the larger coils, traverses that coil and returns by

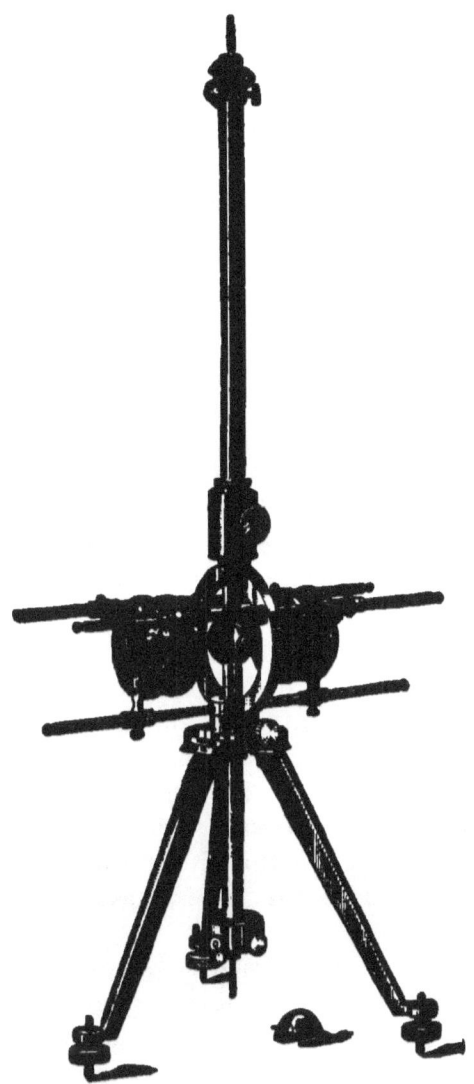

FIG. 61.

another rod to the other large coil, and traversing it, passes out by a binding-screw to the generator through the other terminal of the external circuit.

Proper insulation and connections are provided between the rods, coils, and supporting ring to insure the passage of the current as above; and its direction may be reversed by reversing the connection with the external circuit.

The current in the three coils has practically the same E. M. F., but the difference in resistance gives the high-resistance coil small current strength and the low-resistance coils large current strength, so that the current of the small coil represents chiefly E. M. F., and that of the larger coils, current strength.

When the current passes, its combined effect in the three coils, as represented by the product of the small current into the large, or E. M. F. into current strength, tends to bring the plane of the small coil into a position parallel to that of the other two; the amount of deflection being indicated on a scale by a ray of light reflected from the little mirror, and observed through the aperture shown just above the ring. As this deflection represents the product of the E. M. F. into the current strength, the voltage into the amperage, it shows the electric power of the current as indicated in watts; hence the instrument is appropriately named *electro-dynamometer* or *electric-power-measurer*. It can be used either with the direct or the alternating current, and is especially adapted to the latter, having no magnetic needle.

Measurement of Electric Resistance.—Since current strength depends on the mutual relations of electromotive force and resistance, it is evident that apparatus for varying resistance by the introduction or withdrawal of a definite known quantity, and of ascertaining and

ELECTRIC MEASUREMENT. 155

measuring it when unknown, in order to properly adjust these mutual relations, is a matter of the highest importance in electrical construction. Resistance may be varied, as already shown, by varying the length or diameter of the conductor, or by changing the circuit from series to parallel or the reverse; but as this usually requires permanent construction, it becomes necessary to have also some simple means by which a resistance of known amount can be promptly introduced into any circuit or withdrawn from it without interference with the permanent construction: this is furnished by the *resistance coil*, or rheostat as it is also termed.

Resistance Coils.—Resistance coils are made of german-silver wire on account of its high resistance, which is usually about seventeen times that of pure copper, and calibrated as to gauge and length for a given number of ohms resistance, the wire being properly wrapped for insulation. Fig. 62 gives an ideal view of the construction.

X, Y, and Z are short blocks of brass, insulated from each other above, but connected below through the coils c and d, as shown; each coil being wound with a double strand to reduce self-induction. Two brass plugs, a and b, having hard-rubber handles, fit into holes between the blocks so that when placed as shown, the three blocks are in electric connection, and having practically no resistance, a current would pass directly through them, without traversing the coils. But if a plug, as a, is removed, the current between X and Y must then pass through the coil c. In like manner if plug b is removed, the current between

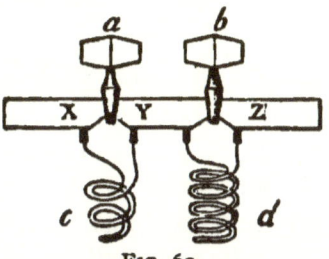

FIG. 62.

156 DYNAMIC ELECTRICITY AND MAGNETISM.

Y and Z must pass through the coil d; which, being twice the length of c, would have twice the resistance if made of wire of the same gauge, or four times the resistance if also the cross-section of the wire were one half that of c. In this way resistance can be varied to any practical extent required.

Sets of resistance coils, calibrated for resistances varying from 1 ohm or less to 10,000 or more, are conveniently arranged in cases, as shown in Fig. 63. The

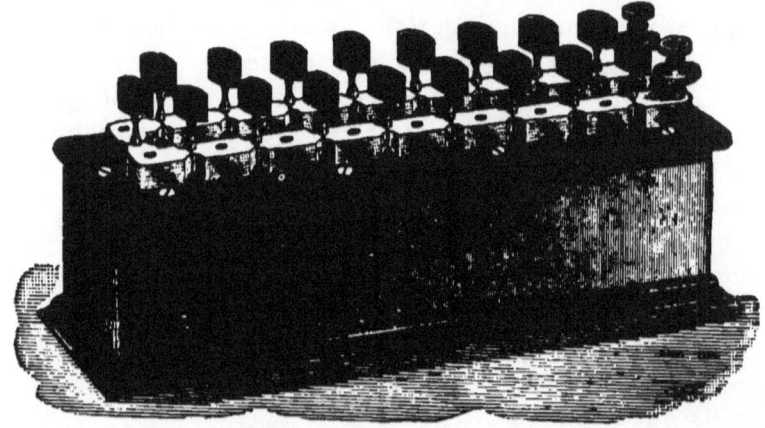

FIG. 63.

case has a hard-rubber cover by which the brass blocks are insulated above, each pair being connected through a coil below, as shown in Fig. 62. A hole in the centre of each block receives each plug when removed from between the blocks, to prevent its being mislaid, and connection with the electric circuit is made through the binding-posts shown at the right.

To introduce any required resistance it is only necessary to remove the plug from its place between the blocks opposite which the resistance required is marked on the cover, the other plugs all remaining connected. If, for instance, 1 ohm resistance is to be introduced, let the first plug at the front right-hand corner be removed, opposite which " 1 ohm" is marked; the current must now flow through that coil, and pass by all the other coils, through the blocks and plugs; if 50 more ohms are to be added, the last plug at the rear left-hand corner is removed, opposite which is marked " 50 ohms;" and the resistance then becomes 51 ohms.

The Wheatstone Bridge.—The Wheatstone bridge is an instrument for measuring an unknown resistance by comparison with a known resistance. Fig. 64 gives an ideal view of its construction. Let A, B, C, D be four wires connected at the points P, Q, M, N, and let M and N be connected with the galvanometer G, and P and Q with the battery X, by which a current can be sent from P to Q. This current will divide at P, and the portion

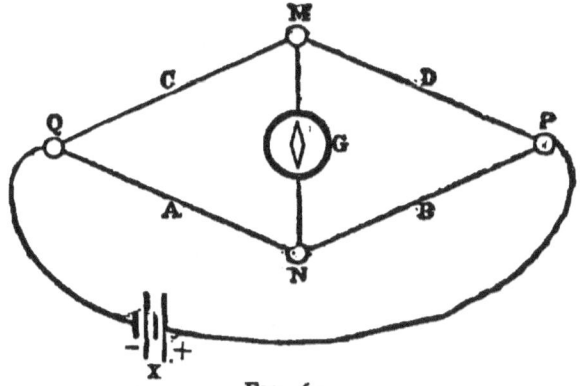

FIG. 64.

passing through each branch of the circuit will be inversely proportional to the respective resistance of each.

Now it is found that the potential between any two points in an electric circuit varies inversely as the resistance between them; and as the E. M. F. between any two points is represented by their potential difference, the E. M. F. at M would vary as the ratio of resistance in C to that in D, and the E. M. F. at N as the ratio of resistance in A to that in B; if these ratios are equal, then the E. M. F., or electric pressure at M, is equal to that at N, irrespective of the amount of current in each branch, and no current can pass between these points, and hence there can be no deflection of the galvanometer needle. But if either ratio differs from the other, then current will pass between M and N in proportion to this difference and produce deflection.

Suppose this difference to be caused by the introduction of an unknown resistance into the arm D; then by varying the resistance in B till the deflection disappears, equality between the ratios is restored, and as the resistances of A, B, and C are known, that of D may be computed; for, allowing the letters to represent the resistances,

$$\text{Since } C : D :: A : B, \quad AD = BC, \quad \text{and} \quad D = \frac{BC}{A}.$$

In like manner, when the respective resistances of any three of the arms are known, that of the fourth may be ascertained.

The total resistance or total current in either branch, or the equality or inequality of resistance or current in the arms, are matters of indifference, equality of *ratios*, as above, being the principle of construction.

As the potential decreases from P to Q in both branches of the circuit, it is evident that if an unknown resistance greater than that of D were substituted for D's resistance, the effect would be to reduce the potential difference, or E. M. F., between C and D, producing deflection of the needle by a flow of current from N to M, and requiring proportional increase of resistance in B to restore the equilibrium. But if this unknown resistance were less than that of D, the effect would be to increase the potential difference between C and D,

ELECTRIC MEASUREMENT. 159

producing deflection by a flow of current from M to N, and requiring proportional decrease of resistance in B.

This instrument may be constructed in any convenient form in which the mutual relations of the different parts to each other are properly maintained; and sets of resistance coils may be so connected with the different arms as to vary the resistance as required. Fig. 65 shows a convenient, practical form.

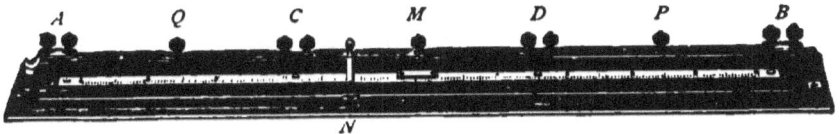

FIG. 65.

On an insulating strip of hard rubber are mounted five copper strips furnished with binding-screws; and between the two end strips is stretched a wire, connected with them, made of a compound metal composed of 85 parts platinum and 15 parts iridium, having high resistance and not easily oxidized; and parallel to it is a graduated scale on which the resistances of equal divisions of the wire are marked in ohms, after proper calibration. The arms and connections for the battery and galvanometer are lettered in the cut to correspond to the lettering in Fig. 64. The arm A extends from Q round to N, including a section of the resistance wire, and the arm B from P round to N, including the remaining section; arm C, from Q to M, and arm D, from M to P: the battery connections being at P and Q, and the galvanometer connections at M and N. The connection at N is made with a slide, mounted on the resistance wire, to which is attached a pointer which indicates on the scale the amount of resistance included in each of the arms A and B. The unknown resistance which is to be measured can be inserted either at C or

D, as preferred, the remaining space being then filled with a known resistance.

When deflection of the needle is produced by the insertion of an unknown resistance at either of those points, a movement of the slide, either to the right or left as required, changes the relative resistances of the arms A and B, and restores the equilibrium by making the ratio of resistance between A and B equal to that between C and D; and the value of the former ratio being indicated on the scale, the value of the unknown resistance can be ascertained, as already explained.

Keys are provided in the battery and galvanometer circuits by which each circuit can be opened or closed as required; the battery circuit being always closed first and opened last, to avoid the violent oscillation of the needle due to the extra current produced by self-induction on opening or closing a circuit.

Fig. 66 shows a very elaborate instrument, combining the galvanometer and a set of resistance coils, by which resistances from one hundredth of an ohm to a million ohms or more can be measured.

The resistance to be measured is connected with the two binding-posts on the left, the battery with the two on the right. Resistance coils ranging from 0 to 10,000 ohms are arranged in four rows of ten each, marked respectively " units," " tens," " hundreds," and " thousands;" and in front of the galvanometer are two rows, A and B, of three each, the corresponding ones on each side marked respectively " 10," " 100," and " 1000."

In the long rows, each of the ten coils in the same row has the same resistance; each in units' row having one unit, each in tens' row one ten, and so on. But in the short rows, each coil has the resistance marked on its bolt. The coils in each long row are connected together in series by the bolts, each coil being connected

ELECTRIC MEASUREMENT. 161

with two bolts by its opposite ends. Parallel to each row of bolts and insulated from them is a brass bar,

FIG. 66.

having practically no resistance; and each of the three bars, marked "units," "tens," and "hundreds," is electrically connected underneath, at the left, to the row of bolts in front of it by the bolts marked o.

When plugs are placed in each of the four holes at the left, opposite the bolts in the four long rows marked o, the current passes directly through the four bolts, plugs, and ends of the bars thus connected, without passing through any of the coils; but if a plug is removed to the right, then the current must pass through all the coils to the left of it in that row and introduce the resistance indicated by the *number* on the bolt and the *word* on the connected bar in front of it. For instance, if a plug connects units' bar with bolt 4, as shown, the current passes through coils 1, 2, 3, and 4, introducing four units of resistance; in like manner the plug connecting bolt 6 with tens' bar introduces 6 tens, bolt three connected with hundreds' bar 3 hundreds, and bolt 7 connected with thousands' bar 7 thousands, making the entire resistance introduced 7364 when the plugs in the two short rows are both opposite bolts numbered alike, as shown.

The two bars parallel to the two short rows are connected underneath by a wire, and each coil in each row has a separate connection with the electric circuit; the three in row A being separately connected at the same point with the arm corresponding to A in Fig. 64, and the three in row B with the arm corresponding to B.

The four long coils connect with the arm corresponding to D; and the resistance to be measured, with the arm corresponding to C. Hence if the resistance in A equals that in B, and the plugs in the four long rows are moved to the right or left till the needle shows no deflection, then the resistance in the four rows must equal that to be measured, since $A : B : : D : C$. Hence, with the plugs placed as shown, that resistance would be 7364 ohms.

ELECTRIC MEASUREMENT.   163

But if a greater resistance than any represented by the four long rows is to be measured, as 100,000 ohms or more, then by changing the plug in row B to bolt 10, and that in row A to bolt 1000, the resistance of arm A is made 100 times that of arm B; hence when the plugs in the four long rows are moved till the needle shows no deflection, the resistance to be measured must be 100 times that indicated in the four rows, which in the special case given would be 736,400. But if the plug in row A were at 100 and that in row B at 10, then, the resistance of A being only ten times that of B, the resistance in the above case, when the deflection was eliminated, would be 73,640.

If a smaller resistance than any represented in the four rows is to be measured, as $\frac{1}{10}$ of an ohm, then by placing the plug in units row opposite 1, and those in the other three long rows opposite 0 in each, and moving the plug in row A to 10 and that in row B to 100, the resistance in A is made $\frac{1}{10}$ of that in B; hence if the needle shows no deflection, the resistance to be measured is shown to be $\frac{1}{10}$ of an ohm. In a similar manner, a resistance of $\frac{1}{100}$ of an ohm may be measured.

Hence we see that when the indicated resistance in row B is greater than in row A, the effect is to divide the indicated resistance in the four rows by the ratio of B to A; but when the indicated resistance in B is less than that in A, the effect is to multiply the indicated resistance in the four rows by the ratio of A to B. In a similar manner any of the indicated resistances can be multiplied or divided.

If, in the construction, the relative positions of arms C and D are reversed, the effect is to reverse the relative positions of arms A and B with reference to them; and hence the multiplication and division, as above.

By increasing the number of coils, and range of resistance, in both the long and short rows, within practical limits, any required resistance, great or small, can be accurately measured.

The battery and galvanometer keys, marked respectively B and G, are shown in front. In a recent form of this instrument the battery key is placed above the galvanometer key and insulated from it, so that the same pressure closes both, the battery key first, as required; and the binding-posts for the battery are placed at the right of the galvanometer, and those for the resistance to be measured at the left; a units' coil is also added to each of the short rows.

In another form of this instrument, the bars are omitted and the resistance introduced by removing plugs, as shown in Figs. 62 and 63.

The plugs should always be pressed in tight, to insure perfect contact.

CHAPTER VII.

THE DYNAMO AND MOTOR.

The Magneto-Electric Generator.—It has been shown in Chapter V that transient electric currents are generated in a conductor forming a closed circuit, when moved through a magnetic field in such a manner as to cut a varying number of lines of force and produce a difference of potential between different parts of the circuit; and that the E. M. F. varies as the number of lines cut per unit of time, and the strength of the current as the E. M. F. divided by the resistance. It has also been shown that when such a conductor is in the form of a coil having a soft iron core, the electric development is greatly increased by the coefficient of magnetism induced in the core.

On these principles the little instrument known as the magneto-electric machine was invented by Pixii in 1833, in which subsequent improvements were made by Saxton and Clarke. It consists, as now constructed, of a short U electromagnet, mounted on an axis, with its poles close to those of a permanent magnet and at right angles to them, and made to rotate rapidly by means of a crank, band-wheel, and gearing. At each make and break thus produced, transient, alternating currents are generated in the coils; and the coil terminals being attached to two brass plates fitted to opposite sides of the axis, with insulating material between them, the currents are taken up and passed to an external circuit by two brass springs which press against these plates.

Commutation.—The plates being insulated from each

other, and out of contact with the springs during the break, and brought into reversed contact with them at the instant of current reversal, which occurs at each half revolution, their position with reference to the springs is reversed as the currents are reversed, and hence the currents are all made to flow in the same direction through the external circuit. A direct current made up of these transient, alternating currents is thus produced by commutation, with perceptible intermission at each make and break, its smoothness varying with the rapidity of the rotation.

Improved machines of this kind were constructed by Siemens, Wilde, and others, among which was a very

FIG. 67.

powerful one, made by the Compagnie l'Alliance of Paris, of the following construction, illustrated by Fig. 67.

The Alliance Machine.—Six bronze wheels, mounted on a horizontal shaft, carried 16 electromagnets on each circumference, 96 in all, which rotated between 7 sets

THE DYNAMO AND MOTOR. 167

of laminated steel magnets, 8 in a set, fixed radially, poles inward, in 8 rows, on a horizontal frame, opposite poles alternating both radially and lengthwise; so that the core of each bobbin, as it rotated between them, was alternately exposed to opposite poles at each end, 16 times at each rotation, the 96 electromagnets thus generating $16 \times 96 = 1536$ transient currents; and as the shaft rotated 350 times per minute, $350 \times 1536 = 537,600$ currents per minute were generated.

A machine with alternating current was employed for the electric light, for lighthouses, and one with direct current for electro-plating and similar work.

The Siemens Armature.—The principal improvement made by Siemens consisted in a new style of bobbin, or *armature*, as it was called, illustrated by Fig. 68, invented in 1856, in which the coils were wound lengthwise, parallel to the axis of rotation, on the flat central part of a long iron core between two flanges, each convex outside and straight inside, and projecting beyond the central part at the ends as shown; a cross-section resembling the letter H.

This armature rotated between large pole-pieces attached to the poles of a powerful laminated steel magnet, the two flanges being the armature's poles, and its coils cutting across the lines of force; and being more fully exposed in the magnetic field than in the old style of winding, the electric development was proportionally increased.

FIG. 68.

Wilde's Machine.—Wilde's improvement consisted in substituting a pair of electromagnets for the steel magnet to produce the magnetic field, and exciting them by

168 DYNAMIC ELECTRICITY AND MAGNETISM.

a small Siemens machine, mounted above it as shown in Fig. 69 ; the Siemens armature being used below as well as above. The current from the armature of the exciting machine passed in circuit through the coils of the electromagnets, while that from the lower armature

FIG. 69.

passed out through the external circuit, being made direct by commutation in both machines. The pole-pieces referred to are indicated in the cut by *m n* above and *T T* below, and insulated from each other by brass indicated by *o* and *i*.

The Dynamo.—Iron when magnetized always retains a little residual magnetism, and when wrought into any

THE DYNAMO AND MOTOR. 169

form acquires a similar quantity by the manipulation. It was proposed by Siemens and Wheatstone, in 1867, to excite the generator by the multiplication of this residual, found in the cores of the electromagnets and armature, by connecting the electromagnet coils with the armature circuit, and thus dispense with the exciting machine. The method of doing this may be illustrated as follows:

In Fig. 69, the magnet coils are connected together below, and have their terminals at p and q above; if the exciting machine be removed and one of the circuit terminals below, as that on the right, be connected with the coils at q, and the other, after passing through the external circuit, be connected at p, then a current passing from the armature out through the left-hand terminal, and traversing the circuit, must return to the right-hand terminal by way of p and q, through the magnet coils, and thence through the armature coils to the left-hand terminal.

The armature of a new machine, so constructed, being put in rotation for the first time, the incipient current generated in its coils during the first half-revolution, by the residual magnetism of the cores, passing through the magnet coils as above, increases this residual, which by its reaction increases the current in the armature coils in like ratio. At the next half-revolution these increased effects are doubled by the mutual reaction; and this doubling occurring at each subsequent half-revolution and being repeated several thousand times per minute by the rapid rotation of the armature, the current, thus continually increasing in geometrical ratio, rises in a few moments to its full normal force, limited by the magnetic saturation of the cores and the carrying capacity of the coils.

The machine, constructed on these principles, was

designated as the *dynamo-electric*, in distinction from the magneto-electric, and subsequently became known briefly as the *dynamo*.

The electromagnets producing the field were called the *field-magnets*, in distinction from the armature, which is also an electromagnet. The springs for taking up the current were called the *brushes*; each consisting of a number of thin copper plates projecting beyond each other at the contact end and soldered together at the outer end. And the pair of insulated segments with which they made contact, and by which the commutation was produced, was called the *commutator*.

Hence the essential parts of the direct-current dynamo became known as the *armature*, the *field-magnets*, the *commutator*, and the *brushes*.

Ladd's Machine.—The current of the machine first constructed by Siemens, in 1867, alternated automatically between the internal and external circuits, being diverted from the latter when employed to excite the former. During the same year a machine was constructed by Ladd, in which the current through both circuits was made continuous. It was substantially the same as the Wilde, with the steel magnet removed, the two armatures retained, one being connected with the magnet coils and the other with the external circuit, and the magnets placed in a horizontal position between armatures of equal size, and supported at each end on large vertical pole-pieces.

The Pacinotti-Gramme Armature.—An armature having the form of a wide ring was invented by Pacinotti in 1862, in which the coils were wound between projections on an iron core. An improvement on this was made by Gramme in 1870, illustrated by Fig. 70, in which the core was composed of annealed iron wires and entirely covered with the coils, only a few of which are shown in the cut; the winding being continuous from coil to coil as shown.

The covering of the core in this manner does not materially obstruct the transmission of magnetic force, copper being diamagnetic, so that such a core is prac-

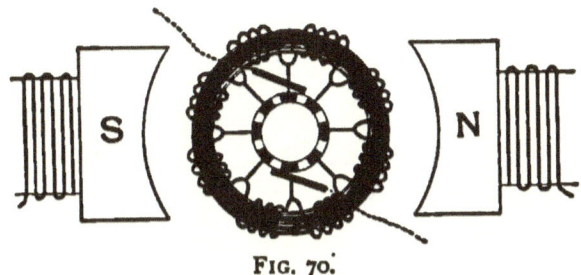

FIG. 70.

tically as susceptible of magnetism as that of the Siemens armature.

Improved Commutator.—An improved style of commutator was also invented, and used by Gramme in the construction of his dynamo in 1870, in connection with his improved armature. It is shown in cross-section in Fig. 70, and consisted of a number of short copper bars mounted on one end of the armature's axis, parallel to its length, and insulated from it and from each other by wood or other insulating material, filling the spaces between them and forming a cylinder under them on the axis. Each bar is attached to a coil as shown, so that the number of coils and bars is equal.

As the currents reverse at each half revolution, a commutator having but two segments produces an intermittent current, as has been shown; but if it have four segments, as shown in Fig. 71, the brushes are brought into contact with two of the segments at each quarter revolution, and if each brush make contact with the approaching segment before breaking contact with the

172 *DYNAMIC ELECTRICITY AND MAGNETISM.*

receding segment, so as to bridge the intervening space, no intermission can occur.

But as the coils, at each revolution, cut a varying number of lines of force per unit of time in different parts of the field, each alternate current must rise with the increase and fall with the decrease of magnetic force; hence, with only four segments, the current,

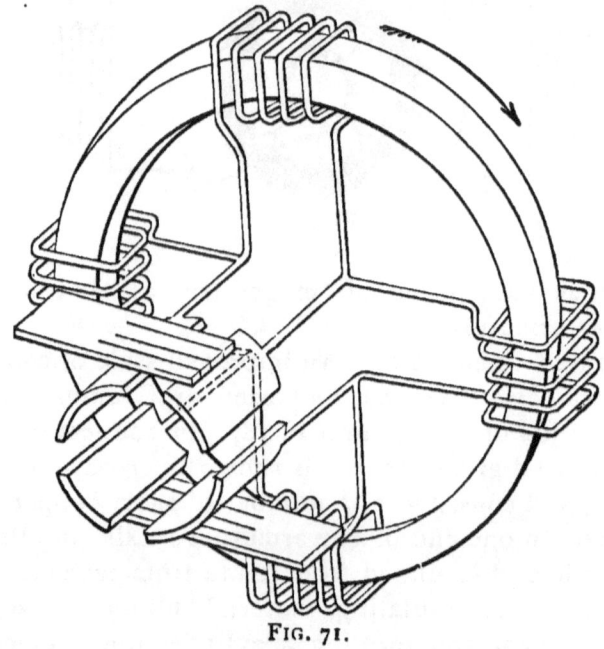

FIG. 71.

though continuous, would be uneven, but with eight segments, as shown in Fig. 70, it becomes at a high speed of the armature practically even, being made still more even as the number of segments is increased.

It is evident that the current cannot pass from one segment to another without traversing all the convolutions of the intervening coil; and as each convolution adds its quota to the current, and each coil is connected with the adjoining coil, all the currents thus generated

THE DYNAMO AND MOTOR.   173

combine to augment the volume of current flowing through the outer circuit.

Direction of the Current.—If the armature, shown in Fig. 70, rotated in the direction of watch-hands and the current, transmitted from it through the field-magnet coils, should circulate in such direction as to induce, in their cores, a north pole on the right of the armature and a south pole on the left. Then, according to the principles of electromagnetic induction explained in Chapter V, the currents generated on the outside part of the right-hand coils of the armature, between its core and the north field-magnet pole, would flow from the observer and be conducted back oppositely through the inside part, while those generated in the left-hand coils would flow in reverse order. And these currents, collected and made direct by the commutator, would enter the external circuit and field-magnet coils by the upper brush, and return to the armature by the lower brush.

If the rotation of the armature were reversed, the direction of the current and polarity of the magnets would be reversed also.

Interior Wire of the Gramme Armature.—Iron being paramagnetic, the lines of force in the magnetic field cannot penetrate the Gramme armature core and pass through the interior of the ring, but are taken up by the core, which thus becomes magnetized. Hence the interior and end wire of the coils does not cut those lines, and cannot in this manner take part in the electric generation, but serves as a conductor of the currents generated in the exterior wire. It also increases the electric generation by the coefficient of magnetic induction received from the core.

According to a theory now somewhat obsolete, the currents are generated by the lines of force threading through the coils, the interior wire thus taking part in the generation equally with the exterior; but experiment seems to prove that this theory is fallacious, as no current is found in the interior and end wire when not continuous with the exterior.

In the Sperry dynamo, interior pole-pieces, parallel to the axis of rotation, are used to render this wire active, the armature rotating between them and similar exterior pole-pieces projecting from the field-magnets.

Another common form of construction is to wind all the wire on the exterior, passing it around projections on each end of the ring.

The Cylinder Armature.—The drum or cylinder armature is also a common form, in which the wire is wound lengthwise on a cylinder, passing over the ends, as shown in Fig. 72. The core generally consists of a large number of thin sheet-iron disks, one of which is shown at B, mounted on a shaft and insulated from each other by tissue-paper. These are usually perforated by openings which, when placed opposite each other, form tubes for interior ventilation, connecting with ventilating spaces between groups of disks, as in the Weston armature, shown at A, on which are also projections between which the wire is wound. They are also made without openings or projections, as in the Edison armature, shown at C, the wire being confined by brass bands, as shown.

This construction of the core prevents the formation of the Foucault currents to which solid cores are liable, and which heat them and serve no useful purpose. And the disks, being parallel to the lines of force and at right angles to the currents, are in the best position for electromagnetic induction. Armatures of the Gramme pattern are also constructed with cores of this kind, made up of flat rings instead of disks.

THE DYNAMO AND MOTOR. 175

The core should come as close to the pole-pieces as possible, to insure maximum magnetic induction, and

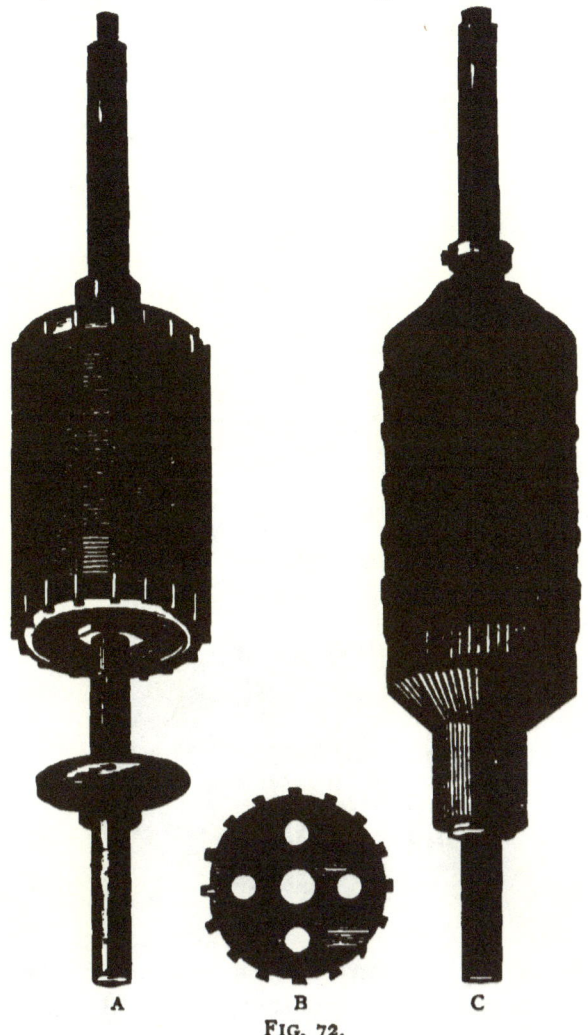

FIG. 72.

hence the wire wound on it should be evenly distributed, and of the minimum quantity and gauge requisite for proper electric induction and resistance.

Closed-Circuit and Open-Circuit Armatures.—Armatures wound like the Gramme, in an endless spiral, with attachment to the commutator segments by radial arms, at regular intervals, are known as *closed-circuit* armatures; and the same designation is applied to those in which the coils are wound separately but connected with each other at the commutator, as in the armature of the Weston dynamo, shown in Fig. 73.

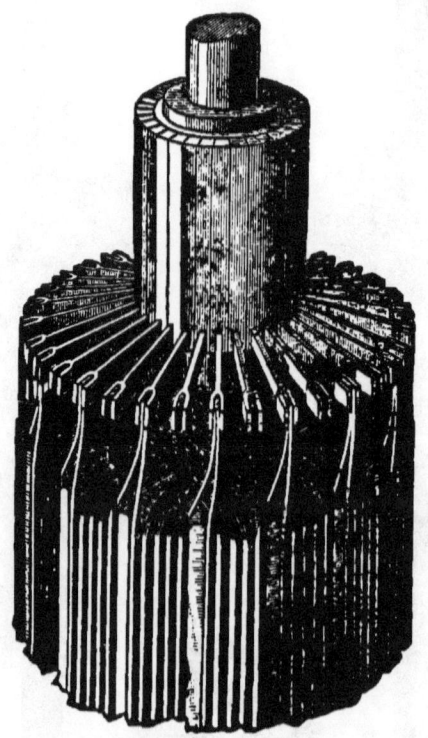

FIG. 73.

In another style, known as the *open-circuit* armature, each coil is independent of every other, its terminals being connected to two opposite segments of the commutator which have no connection with the other coils,

THE DYNAMO AND MOTOR. 177

as in the armature of the Brush dynamo; hence only those coils connected with the brushes through the commutator are in action simultaneously, each set coming into action as the other set passes out. Four brushes are employed in an eight-coil Brush dynamo, and the contacts are made in such a manner that six coils are in action simultaneously.

Location of the Armature's Magnetic Poles.—In accordance with the principles of magnetic induction, the polarity induced in the core of the armature by the field-magnets during rotation is opposite to that of the inducing poles, as shown in Fig. 74. But this polarity

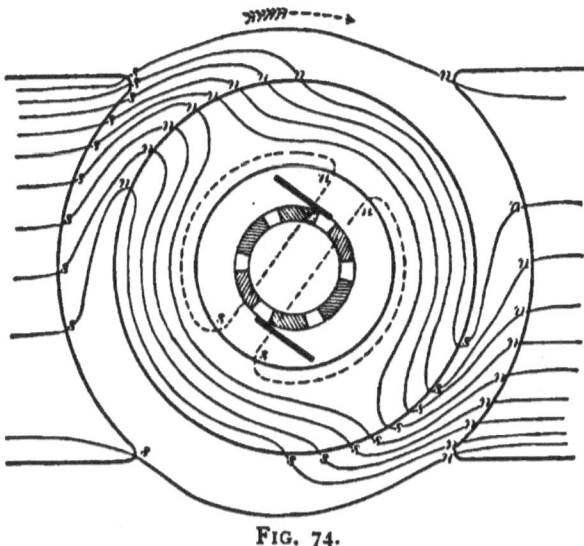

FIG. 74.

is comparatively weak, the core's most effective polarity being that induced by the currents circulating through the armature's coils, the tendency of which is to induce similar poles in proximity to those of the field-magnets which, by mutual repulsion, are deflected into the position indicated by *n n* and *s s* on a line joining the brush contacts; each half of the core, divided on this line, becoming a separate magnet.

The poles of the field-magnets are deflected in the opposite direction, the north pole to the lower corner of the pole-piece on the right, and the south pole to the upper corner of the pole-piece on the left; a line joining their centres being nearly at right angles to that joining the stronger armature poles. Hence the lines of force become contorted as shown.

Magnetic Lag.—The armature core does not become fully magnetized at the instant induction occurs, nor fully demagnetized at the instant it ceases; an infinitesimal moment being required for its saturation in the first instance and its demagnetization in the second, known as *magnetic lag*, during which its poles are carried slightly forward in the direction of the rotation; this tends to separate the dissimilar poles induced by the field-magnets from the field-magnet poles, and thus to increase the contortion of the lines.

Position of the Brushes.—The brushes make contact with the commutator on or near the neutral line on which the currents reverse, as shown in Fig. 74, and where consequently no currents are generated; hence, in a closed-circuit armature, the parallel currents generated on the left pass out from the armature by the upper brush, as each segment of the commutator comes into contact with it, and those generated on the right are added to the inflowing current entering the armature by the lower brush.

If the brushes were shifted into the line of highest potential, which is at right angles to the neutral line, the wire in which the parallel currents are generated, on either side, would be carried round by rotation to the opposite side before the connecting commutator segments reached the brush, and the currents neutralized by opposing currents generated in the wire, and the external current cease.

But if the brushes made contact on a line between the neutral line and line of highest potential, a partial neutralization by opposing currents would occur, and the electric potential vary as the distance of the brushes from the neutral line; increasing as they approached it and decreasing as they receded from it. By shifting the brushes in this manner, automatically or otherwise, the potential and resulting current can be varied and regulated as required.

Such regulation is common, but its range is limited, and it cannot always be used advantageously, as it tends to increase sparking at the brushes, a wasteful and injurious heating effect, difficult to suppress entirely.

The Field-Magnets.—The field-magnets of different dynamos vary greatly in construction and constitute the principal part of the framework of each machine, and hence they are so constructed as to support the various parts in the most convenient manner and give a compact, appropriate form, without interference with their special function.

They have massive cores, usually of the best cast-iron, preferably annealed, malleable iron, though wrought-iron is also employed, but the advantage is not usually sufficient to compensate the extra cost. These cores should be sufficiently massive to insure the absorption of all the magnetism which can be generated in them without over-saturation. They terminate, at one end, in enlarged pole-pieces which nearly inclose the armature, the opposite ends being connected by a cast-iron yoke, or bolted together by cross-bars, to complete the magnetic circuit. They are wound with heavy insulated copper wire, the winding being continuous from core to core.

180 *DYNAMIC ELECTRICITY AND MAGNETISM.*

A single pair of such magnets may be employed, or two or more pairs, each core having a separate pole-piece, or two or more cores being joined to the same pole-piece.

Series, Shunt, and Compound Winding.—There are three principal methods of winding the field-magnets, known respectively as the *series*, the *shunt*, and the *compound* winding.

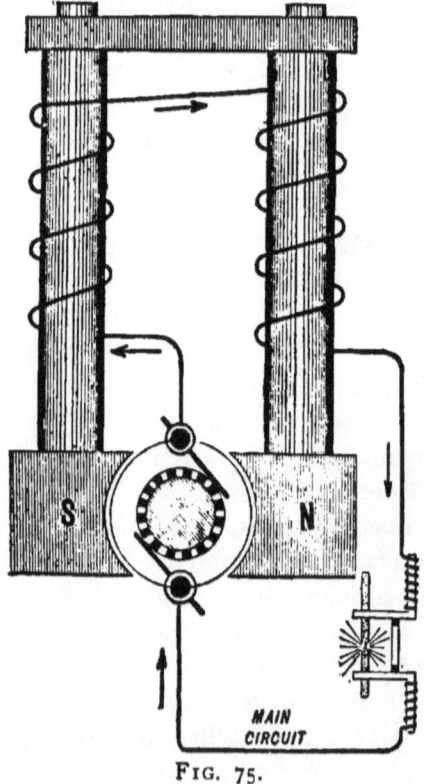

FIG. 75.

In the series method, illustrated by Fig. 75, the entire current traverses a single route of low resistance, passing in series through the armature, the field-magnets, and the external circuit; so that any variation of resistance, at any point, affects the entire series equally.

In the shunt method, illustrated by Fig. 76, the current traverses two distinct routes; dividing, at the upper brush, in the inverse ratio of the resistance of each circuit. The main current flows to the right through the

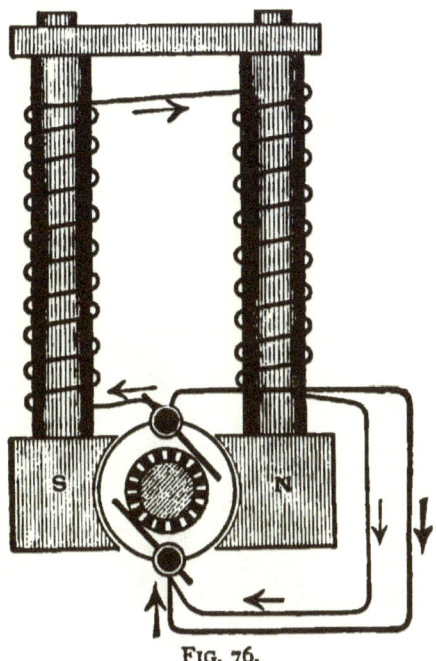

FIG. 76.

coarse wire of the external circuit, while a small current, varying from 1.5% to 20% of the entire volume, flows through the shunt, or fine wire with which the magnets are wound, and is employed exclusively to excite them.

If the resistance of the main circuit is increased, the strength of its current is proportionally diminished. But the potential difference, or E. M. F., between the brushes, representing the electric pressure, is increased by the diminished flow of current in the ratio this increased resistance bears to itself plus the armature's resistance: and as the resistance of the shunt remains constant, the strength of its current is proportionally increased by this increase of E. M. F.: and the magnetism of the core being increased in the inverse ratio of its saturation, by this increase of current strength in the shunt, its reaction increases the current strength in both cir-

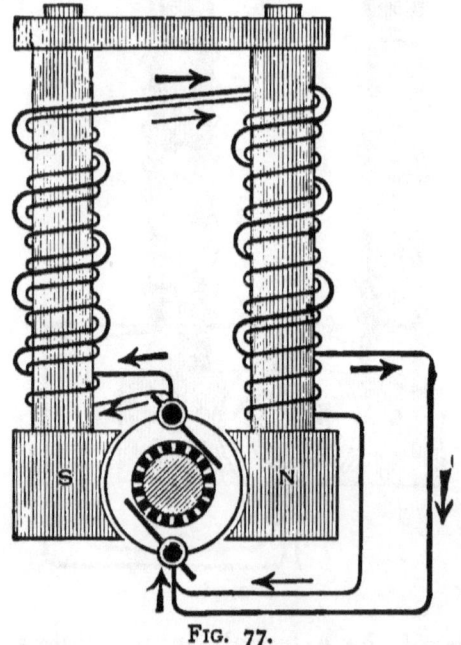

FIG. 77.

cuits; thus supplying electric energy to overcome the increased resistance. By this series of adjustments an equilibrium between these various factors is established, the total electric energy developed, varying as the mechanical energy expended. Decrease of external resistance reverses these results.

The resistance of the shunt may be varied as required, by resistance coils.

The compound winding, illustrated by Fig. 77, is a combination of the series and shunt methods; a shunt wire of high resistance, used only to excite the magnets, being employed in connection with the low resistance wire, which is wound by the series method and excites them also. The automatic regulation is similar to that of the exclusive shunt method, except that the entire current flows through the magnet coils.

Each of these methods of winding has its special adaptation to the requirements of a certain kind of work; as, for instance, in electric lighting it is found that the series-wound machine is usually the most suitable for arc lighting, and the shunt and compound wound for incandescent lighting; arc lighting requiring high E. M. F. and comparatively small current, while the requirements for incandescent lighting are the reverse; which leads to the classification given below.

Constant Current Dynamo.—To maintain a number of arc-lamps, connected in series, at a given illumination, a constant current of ten or more amperes, flowing from lamp to lamp, is required for each. If but one lamp were lighted, the required E. M. F. or potential would be comparatively small; but if two lamps were lighted, the resistance being doubled, the E. M. F. must be doubled to maintain the same current strength; and the same ratio of E. M. F. to resistance must be maintained for any number lighted or extinguished.

Hence the construction and regulation of a dynamo for this work, or any work having similar requirements, must be such as to furnish E. M. F. capable of variation within the required range; and a machine so constructed is known as a *constant-current* dynamo, and is usually series-wound as stated above.

Constant-Potential Dynamo.—But if the required work were the maintenance at a given illumination of a

184 *DYNAMIC ELECTRICITY AND MAGNETISM.*

number of incandescent lamps connected in parallel, the lamps being on branches derived from the main circuit, the variation of resistance is confined to these branches, in which it becomes adjusted to the requirements of the current, the resistance of the main circuit

FIG. 78.

remaining constant; hence the E. M. F. remains nearly constant; and a machine adapted to such work, or work having similar requirements, is known as a *constant-potential* dynamo, and is either shunt or compound wound.

THE DYNAMO AND MOTOR.   185

The Edison Dynamo.—Dynamos differ greatly in appearance and minor details of construction, but their general construction and the relations of the different parts will be readily understood from the Edison dynamo, shown in Fig. 78, which is a direct-current, shunt-wound machine, used especially for incandescent lighting, and a fair representative of its class.

The field-magnets, mounted vertically, rest on massive pole-pieces inclosing the armature below, and on their left are shown the connections of the coils, the lever above by which the external circuit, represented by the projecting terminals, is connected and disconnected, the projecting end of the armature below, with the commutator and brushes, the latter attached to a yoke, movable manually for adjustment of potential. The oil-cups, band-wheel, and screws for shifting the machine's position, to tighten or loosen the belt, are also shown below, and the lamp above, which indicates the general state of the current.

Alternating Current Dynamos.—The transient currents generated by the armature, when passed into the external circuit without commutation, produce a continuous alternating current, and electromagnetic machines having such construction are known as *alternating current* dynamos.

The Gordon Dynamo.—The older machines of this class have a construction somewhat similar to that of the Alliance magneto-electric machine, already described; electromagnets with alternating poles taking the place of the steel field-magnets. The Gordon machine is of this construction; 64 short field-magnets being mounted transversely on the circumference of a wheel which rotates between two stationary armatures of similar construction, each having 64 coils; and the coils being oppositely wound on each alternate bobbin, both in the

186 DYNAMIC ELECTRICITY AND MAGNETISM.

armature and field-magnets, produce alternating poles in each.

The currents flow from the armature coils to the external circuit without the intervention of a collector and brushes; and the field-magnets are excited by two direct-current dynamos.

The Westinghouse Dynamo.—The Westinghouse alternating-current dynamo represents an improved method

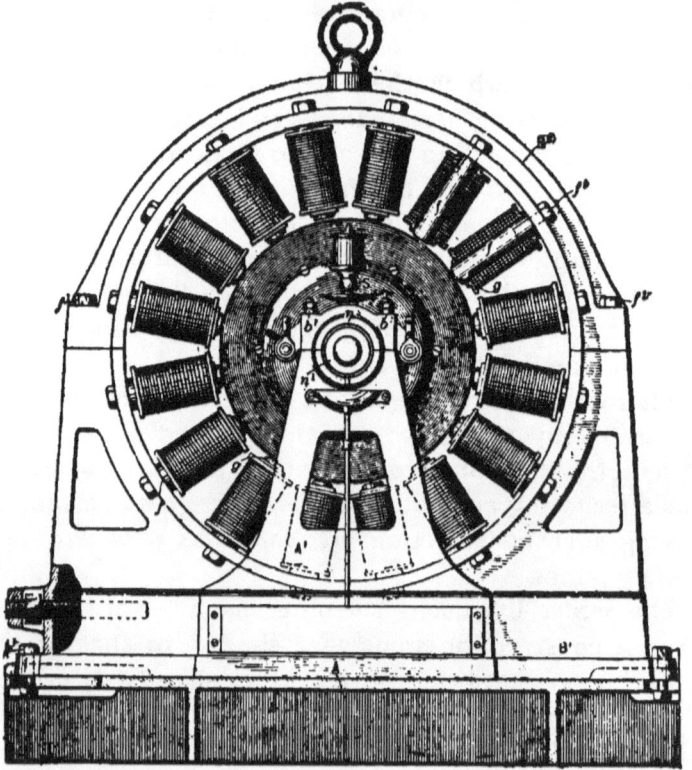

FIG. 79.

of construction, the principal features of which have been adopted by several machines of this class.

Fig. 79 is a sectional view of the machine, as seen from the end of the armature shaft, representing 16

THE DYNAMO AND MOTOR. 187

field-magnets attached radially to a circular frame, their opposite, alternating poles inclosing a central space in which the armature rotates; their cores and winding being shown in section above.

Fig. 80 is a sectional view parallel to the shaft; a side

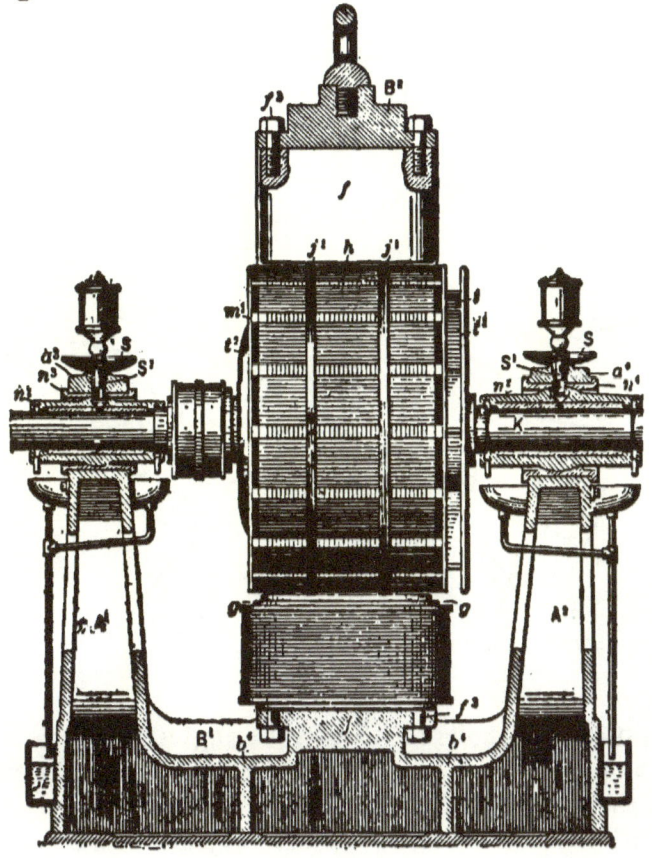

FIG. 80.

view of one of the field-magnets being given below, and that of a core above. Mounted on the shaft at the left of the armature is the collector, composed of two copper rings, insulated from each other, on each of which a

188 DYNAMIC ELECTRICITY AND MAGNETISM.

brush, connected with a separate terminal of the external circuit, makes contact.

The armature core is composed of insulated sheet-iron disks, and ventilated by tubular openings in the manner already described; and the coils are wound in a single layer on its external surface and looped around projections at the ends. The manner of winding is shown in Fig. 81, a correct idea of it being obtained by

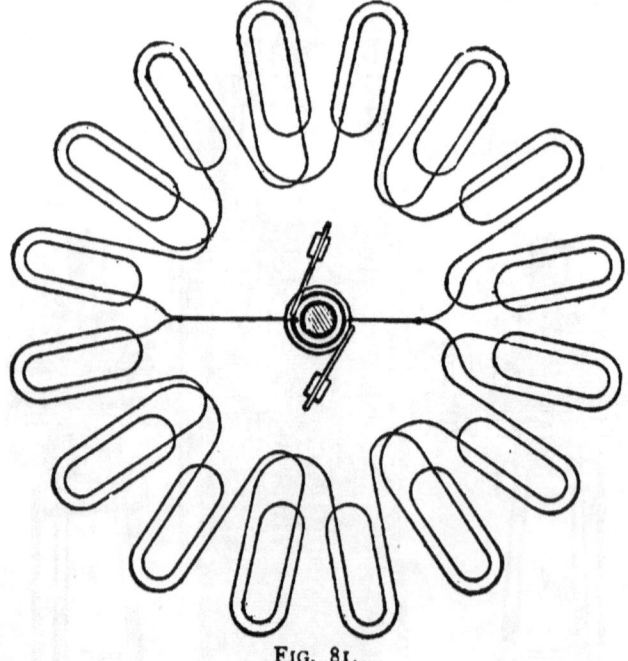

FIG. 81.

supposing the coils to lie at right angles to the surface of the paper, the outer ends turned from the observer and the inner ends towards him. Each alternate coil is oppositely wound as shown, and they all form a continuous closed circuit, the opposite terminals of which are connected with the separate rings of the collector; the current passing out from one ring and returning to the other alternately.

THE DYNAMO AND MOTOR. 189

Separate Excitation.—The direct current, always required for exciting the field-magnets, in the Westinghouse and similar dynamos, is obtained, as in the Gordon, from a separate, small machine. This separate excitation, which involves extra expense, complication, and inconvenience, may be avoided by the generation of a separate, direct current in the machine itself by the commutation, for this purpose, of a small portion of the alternating current. But separate excitation is found to be the most practicable for the large dynamos usually employed for alternating-current work.

Advantages of the Alternating Current Dynamo.—The peculiar construction of the alternating-current dynamo and the elimination from it of the commutator, with its resistance and wasteful sparking, results in the generation of currents of much higher potential, with less internal resistance than it is possible to obtain from the direct-current dynamo. Such currents can overcome the resistance of the external circuit more efficiently than those of low potential; and on this principle is based the practical rule that the amount of copper in the conductor should vary inversely as the square of the E. M. F.; according to which it is found possible to transmit such currents to points remote from the generator by comparatively small wires, and thus distribute electric energy, for practical use, over a much larger area, at the same cost, than is possible with the direct current system; or over the same area at far less cost.

This economical advantage is increased where electricity can be generated more cheaply, as by water-power, at a point remote from where it is required for consumption; or where the generating station can be located on cheap property to furnish current for use on more expensive property, as often happens in cities.

The Converter.—Incandescent lighting is the principal

use for which the alternating current is now employed; and as this requires a large current distributed in small parallel currents among a great number of lamps, as explained in Chapter XI, 5000 being sometimes thus illuminated by a single dynamo, the conditions of high potential and comparatively small current, under which the electric energy is delivered, require to be reversed at the several points where it is to be consumed.

This is done by the apparatus known as the *converter* or *transformer*, which is simply an inverted induction-coil of special construction; the primary coil consisting of fine wire which receives the high potential current from the dynamo, and the secondary coil, insulated from the primary, consisting of coarse wire in which, in consequence of the low resistance, a large current is induced and supplied to the lamps.

Instead of an iron core inclosed by the coils, the coils are inclosed in an iron case composed of insulated sheet-iron plates, built up in the same manner as the armature cores already described; the two coils being placed side by side, so that both are equally exposed to the magnetic induction.

These converters, mounted on poles or otherwise, are distributed along the line between two parallel wires, one connected with the primary coils and dynamo, and the other with the secondary coils and lamps. At each point where light is required, a converter of the capacity requisite to furnish current for the required number of lamps is placed. Ten to eighty lamps may thus be supplied from the same converter.

Development of the Electric Motor. — Oersted's discovery of electromagnetic action, in 1819, and the subsequent development of electromagnetism to which it gave rise, led to the invention of numerous machines designed to utilize electricity as a motive power by means of the electromagnet. The principle of construction in all these machines consisted in energizing electromagnets by a battery current, and by their attraction and repulsion producing mechanical motion, either rotary or oscillating.

In the rotary motors a number of iron armatures, with or without inclosing coils, rotated in proximity to an equal number of stationary electromagnets; the rotation being produced by the attraction of each armature in the same direction by the opposite magnetism of a stationary pole, and its repulsion by the similar magnetism of the pole from which it was receding; the polarity being reversed at the instant of closest proximity by a commutator fixed on the rotating shaft.

The Jacobi motor was among the most noted of the coiled armature class, and the Froment and Neff motors of the naked armature class; the armatures in the Neff being stationary and the magnets rotary, while in the Froment the armatures rotated and the magnets were stationary.

In the oscillating machines the armatures, consisting of a pair of loose fitting pistons, were attracted alternately into hollow electromagnets whose polarity was reversed by a commutator at the close of each oscillation, and a reciprocating motion thus produced. These pistons and magnets were placed either vertically at the opposite ends of a horizontal walking-beam, as in the Gustin motor, or horizontally, end to end, as in the Du Moncel and Page motors, in which the oscillatory motion was changed to rotary by a crank.

From 1830 to 1873 various motors of the above kinds were constructed, and attempts made to operate machinery and propel boats and cars with them; one of the most noted of these experiments having been made by Jacobi with his motor at St. Petersburg in 1838; with which he propelled a boat on the Neva, carrying 14 passengers, at the rate of three miles an hour; employing first a Daniell battery of 320 cells, and subsequently a Grove battery of 138 cells. The Daniell was objectionable on account of its great weight, and the Grove on account of its noxious fumes, while the rate of speed was far too low to be of any practical advantage.

In 1851 Page propelled a car on the Washington and Baltimore Railroad, at a maximum speed of 19 miles an hour, with a 16-horse motor of his construction and a Grove battery of 100 cells.

But the limited capacity of motors constructed as above, the cost and inconvenience of batteries of the requisite size, their want of constancy for such strong currents, and the risk and difficulty of their transportation when filled with fluid and employed to propel cars, were fatal defects which could not be overcome, so that all such motors were found to be impracticable.

In 1861 Pacinotti invented a motor, the armature of which has already been described; the whole construction being practically the same as that of the Gramme dynamo which appeared subsequently. This motor, like all its predecessors, was energized by a battery, and hence could not be made practical, but was the same in principle as the improved motors now in common use; being simply a reversed dynamo in which an electromagnetic current produced mechanical motion, instead of mechanical motion producing an electromagnetic current.

Pacinotti recognized this principle of inversion, having found that by energizing his field-magnets by the battery current, or substituting permanent magnets for them, and rotating his armature mechanically, an electric current was generated; so that the machine could generate motion by applying current, or current by applying motion.

The Dynamo as a Motor.—This principle of inversion in the dynamo, as discovered by Pacinotti, received its first practical application by Fontaine at the Vienna exposition in 1873; when he used a Gramme magneto-electric machine, attached to a pump, as a motor, putting it in operation by a current from a Gramme dynamo. This led to the discovery that the dynamo itself could be used as a motor and operated by a current supplied by another dynamo; thus substituting the stronger, cheaper, constant current generated by mechanical power for the weak, dear, inconstant current generated by chemical action, and the superior energy of electromagnetic action for mere magnetic attraction and repulsion.

Hence the motor and the dynamo are identical in principle and in construction, and the same machine may be used either as a generator or a motor. In practical use, however, the motor usually requires to be smaller and more compact, as the power generated by a steam-engine or a water-wheel can be converted into electricity most economically by a large dynamo, and distributed for running cars or operating light machinery by numerous small motors; a motor of a few pounds weight having sufficient capacity to operate a sewing-machine or a small lathe.

Principles of the Motor.—According to Lenz's law, referred to in Chapter V, the reaction of an induced current, generated by the mechanical movement of a conductor, is always in opposition to the movement; hence the currents induced in the armature of a dynamo react in opposition to its rotary movement. This reaction is the result of the potential difference generated by the movement, which, as has been shown, induces opposite

194   DYNAMIC ELECTRICITY AND MAGNETISM.

electromagnetic poles in the adjacent parts of the armature and field-magnets, producing attraction which tends to arrest the rotation. This attraction is electric as well as magnetic; the currents generated in the coils of opposite poles, facing each other, flowing in the same direction, and hence being mutually attracted.

Now it is evident that a mechanical rotary force equal to this reaction is necessary to overcome it, and this constitutes almost the entire force required; the force requisite to overcome the friction and inertia of the armature being comparatively insignificant; and it is the rotation of the armature in opposition to this reaction which generates the current.

If a dynamo, put in operation in this manner, be connected by conductors with another dynamo intended to act as a motor, the above conditions of potential difference and reaction are produced in the second machine by the current from the first; and there being no mechanical force in the motor to oppose this reaction, its armature rotates in the opposite direction to that of the generator, reproducing the mechanical force applied to the latter, less a certain percentage consumed in overcoming the resistance of the conductors.

Hence the principle of the motor is simply that originally discovered by Oersted, the rotation of a magnet by an electric current.

But the motor thus operated generates a counter-current in opposition to that of the dynamo, and when the two machines are of equal capacity the opposing currents vary as the relative speed of each machine; the motor current increasing and the dynamo current decreasing till the speed is equalized, when the strength of the motor current becomes equal to that of the dynamo current less the amount necessary to overcome the motor's friction and inertia, and no effective current flows from the dynamo.

This condition is soon attained when the machines are running without "load," that is, without doing useful work. But when the motor is made to operate machinery its speed is reduced in proportion to the load, and the counter-current decreasing as the speed, the current from the dynamo is increased in the same ratio.

Hence, as the current varies inversely as the motor's speed and directly as the dynamo's speed, and the motor's speed varies inversely as the load, it follows that the speed of the dynamo must be made to vary directly as the load of the motor in order to maintain the requisite speed in the motor for the performance of useful work. Hence variation of load at the motor requires corresponding variation of power at the dynamo; the combined machines being simply an apparatus for the convenient application of mechanical power to useful work; mechanical energy being transformed into electric energy by the dynamo, and this electric energy transformed into mechanical energy by the motor.

Loss of Energy.—This double transformation entails a loss of about 15 % of the mechanical energy derived from the steam-engine or other source of power; this percentage being spent in overcoming the friction, inertia, electric resistance, and self-induction of the machines, including also their incidental waste. Besides this a loss is incurred in overcoming the electric resistance of the conductors, which varies in proportion to their cross-section and required length, and may equal an additional 10 per cent or more, according to the distance to which the power is to be conveyed. A considerable loss is also often incurred by imperfect insulation, unavoidable under certain conditions, as in the running of street-cars.

Eddy Currents.—In both the dynamo and motor, currents are induced in the iron core of the armature, unless suppressed by specific means, which in the dynamo flow in the same direction as those induced in the coils, and in the motor, in the opposite direction. These currents, regarded by Foucault as magnetic, are regarded by later writers as electric; a distinction which pertains chiefly to their direction rather than their nature, if both kinds of energy be considered identical; and since they cannot combine with the currents in the coils, they serve no useful purpose, circulating as eddies in the iron, wasting energy and generating heat.

The laminated structure of the armature core suppresses them almost entirely in the dynamo, as has been shown, but it has been found more difficult to suppress them in the motor. For in the dynamo the two sets of currents, being in the same direction, tend to weaken each other, while in the motor, being in opposite directions, they tend to strengthen each other, in accordance with the principles of current induction. Hence, in the dynamo the useful currents tend to suppress the eddy currents, and in the motor, to increase them. So that any eddy currents induced in the core of either armature, notwithstanding the lamination, become more prominent in that of the motor.

As these eddy currents are regarded as the chief cause of loss of energy in motors, the importance of suppressing them by complete lamination, with thin disks and perfect insulation, in motor armatures of all sizes, is apparent.

Series, Shunt, and Compound Wound Motors.—The field-magnets of motors, like those of dynamos, are either series, shunt, or compound wound, and machines of each style are applied to the same work; practice being less definitely settled in motor work than in dynamo work, and opinion in regard to the adaptability of the different styles of winding to the different kinds of motor work varying. This arises from the complicated character of an apparatus composed of two machines having opposite functions and reversed modes of action, the adjustment of whose mutual relations, so as to adapt the apparatus to a varying external load, presents a problem far more difficult of solution than the adaptation of a single machine to similar work, as in the dynamo.

The shunt wound motor has the advantage of its internal, automatic regulation, which adapts it to stationary work having approximate constancy of load; while the series wound has been found better adapted to street-car work, where starting and stoppage, varying grade and speed, and varying number of passengers require manual regulation to adapt the current to this varying load on the motor, also prompt reversibility of motion and command of maximum energy at any rate of speed.

In the Sprague motor, which is compound wound, differential regulation is obtained by opposing the shunt to the series current.

In these different styles of winding the current enters, leaves, and circulates through the field in the reverse order to that in which it enters, leaves, and circulates through the corresponding styles of winding in the dynamo, except as changes are required in specific methods of regulation.

Reversible Rotation.—But as reversal of mechanical motion is often desirable, motors are constructed in which the rotation of the armature is made reversible. This is accomplished in a very simple manner with two sets of brushes having opposite current connections; one set being lifted out of contact with the commutator by a lever attached to the brush-yoke, as the other set is brought into contact; and the direction of the current through the armature being thus reversed, the rotation is reversed also.

The Alternating Current Motor.—The construction of the direct current motor involves the double conversion of the electric current from alternating to direct in the dynamo and from direct to alternating in the motor, the armature currents in each machine being alternating and the field-magnet and line currents direct.

As this double conversion requires two commutators with their wasteful resistance and sparking, various attempts have been made to eliminate it and produce a strictly alternating current motor; but previous to 1888 such motors had not been made practically successful. The general introduction of the alternating current system at about that date created a special demand for them, and led to the construction of a practical motor of this kind by Nikola Tesla, based on the principle of the shifting of the magnetic poles in the dynamo, which has already been described.

Tesla made the important discovery that if the field-magnets of a motor were made in the form of a ring, similar in construction to that of the Gramme armature, the coils on opposite sections of the core all around being separately connected in pairs, and the terminals of alternate, opposite pairs connected with two pairs of the usual collectors of the alternating current dynamo, on which the brushes make contact, the magnetic poles, induced in the motor by the transmitted currents, would be shifted continuously round the ring from pole to pole, making a complete revolution during each revolution of the dynamo's armature, the polarity of each section of the ring being reversed by each alternation of current: and opposite poles being induced in the armature of the motor, it would be put in rotation in a corresponding manner by the resulting electromagnetic attraction and repulsion.

The chief difference between this method of rotation and that of the direct current motor consists in the fact that, in the latter, the poles, having shifted to their relative positions, remain stationary, the armature rotating and its poles shifting continuously in the opposite direction as the armature rotates through them. Now if the field-magnets of the Tesla motor be regarded as a stationary armature, and the armature as a rotating field-magnet, we have practically the same relative conditions as in the direct current motor, with this difference, that in the Tesla the poles rotate through a stationary armature, instead of the armature rotating through stationary poles; and the stationary poles of the rotating field-magnet maintain practically the same constancy of relative position to the rotating poles of the stationary armature as is found between the corresponding poles in the direct current motor. The relative conditions in the two motors are reversed but not essentially changed except by the elimination of the commutator from the Tesla. And the ability of the Tesla to start, stop, and maintain its rotation in synchronism with the dynamo is due to the reversed construction of the two machines, by which the magnetic poles are made rotary in the one where stationary in the other, and stationary in the one where rotary in the other.

The Westinghouse Tesla Motor.—A Tesla motor, used in connection with the Westinghouse alternating current dynamo, is shown in Fig. 82, its construction being similar to that of the dynamo with certain modifications. The core of the field-magnets, or stationary armature, is laminated, each plate being circular externally, and having an even number of poles projecting inward. These plates, arranged symmetrically and properly

200 DYNAMIC ELECTRICITY AND MAGNETISM.

insulated, are bolted together between two caps which form the ends of the supporting frame; the lamination being shown on the outside in the cut, and twelve poles through the ventilating openings in the caps. The coils are in two separate series, wound oppositely on alternate poles; those of the same series being all wound in the

FIG. 82.

same direction, and those of the other series, which alternate with them, in the opposite direction.

Each series has one of its terminals connected with the binding-post on the right, the other terminals being each connected with a separate binding-post on the left; the two left-hand posts being connected with separate collecting rings on the dynamo, and the right-hand post with a third ring; hence the current which enters by the right-hand post divides, producing opposite poles in the alternate, oppositely wound coils, and leaves by the two left-hand posts; while, at the next alternation, the current enters by the two left-hand posts, reverses the polarity, and leaves by the right-hand post.

The armature, or rotating field-magnet, has the same construction as the armature of the Westinghouse dynamo, a single layer of copper wire covering the circumference of a laminated iron cylinder, mounted on the shaft, the coils being looped round projections at the ends. The winding is continuous, as in the dynamo, but the two terminals are soldered together, so that the coils form a closed circuit without external connection. The insulation of the armature is not of essential importance, its E. M. F. being very low.

In the direct current motor, with an armature rotating between two pole-pieces, there are only four poles, two in the field-magnets and two in the armature on a line at right angles to them; but in this motor there are 12 alternate poles in the field-magnets and 12 in the armature; each pair of armature poles, on opposite sides, being on a line at right angles to that joining a pair of field-magnet poles on opposite sides; the polarity of each pair being similar but opposite to that of the other pair. Hence the poles rotating round the stationary field-magnets can never be more that $\frac{1}{12}$ of the circumference in advance of the fixed poles which follow them; and there is a tangential pull between the dissimilar poles of each element, in opposite directions at opposite ends of each diameter, and a tangential push between the similar poles, producing the rotary force.

As this attraction and repulsion varies inversely as

the square of the distance, and as the distance between poles varies with their number, it is evident that the rotary force must vary in the same proportion. But it should be distinctly borne in mind that this force is not due to magnetism alone, but is electromagnetic, as in the direct-current motor, the poles of each element being constantly deflected by those of the other element, in accordance with the principle discovered by Oersted.

The mutual relations of speed and current, and the generation of a counter current, explained in connection with direct current motors, apply also to this motor. As the speed, with a given load, varies in proportion to the rotary force, and the force varies with the number of poles, it is evident that the speed must vary as the number of poles. Hence there should always be a sufficient number to insure the necessary speed for the required work, and this number may be greater or less than the number required in the connected dynamo; the relative speed of the two machines varying inversely as the relative number of poles in each. These general principles are, however, subject to various modifications dependent on special modifications of construction in each machine.

The Tesla Motor as a Converter.—The construction of this motor is practically the same as that of the converter—a laminated iron case inclosing two insulated copper coils acting inductively on each other. This construction magnifies the inductive effect by bringing the coils of each element into close proximity, and especially increases the facility for rapid reversal of rotation, which is effected in the following manner:

Reversal of Rotation.—If, while the motor is in operation, the connections of the two left-hand binding-posts with the dynamo be reversed by the movement of a switch, the polarity will be reversed and the poles of the field-magnets made to rotate in the opposite direction. This reversal of polarity in opposition to the momentum of the armature changes the motor for an instant to a dynamo, generating a strong extra current in the armature coils. This is the extra current of break, described in connection with the induction-coil as having superior strength, and, its direction being the same as if there had been no reversal, it opposes for an instant the reversed current from the dynamo, arresting the rotation of the armature, which, at the next instant, is reversed by the reversal of polarity; the successive steps following each other so rapidly that full speed in one direction is almost instantly changed to full speed in the opposite direction.

Similar effects are produced in the reversal of the direct current motor, but the construction does not admit of equally effective induction.

By passing the current through a special converter, the regulation of this motor can be adapted to any practical conditions required.

Distribution of Power.—A number of motors may be operated on cars, or in shops, from one or more large dynamos, coupled together as generators, or employed separately, and centrally located; the parallel system of distribution, described in Chapter XI, being usually adopted, having been found more practical than the series system.

This system is practically the same for distribution of power as for distribution of light, but in its application to cars special methods are required. One of the most common is to suspend a wire above the track and make connection between it and the motor on each car by a trolley attached to the end of a connecting-rod which projects above the car. As the direct current is usually employed, it enters by this wire, insulated in the air, and returns by the rails, which are electrically connected together for this purpose; the motor having electric connection with them through the car axle and wheels, and the dynamo similar direct connection.

As the resistance between the mains varies inversely as the number of cars employing current simultaneously, the supply of current varies in like proportion; self-regulation being thus obtained, as in parallel lighting; any additional regulation required being supplied by resistance coils or otherwise.

In cities where municipal regulations prohibit air lines, both conductors may be placed in a conduit; but as it is difficult, in such construction, to maintain proper insulation for naked wires with the open slot required for connection with the motor, various ingenious methods have been devised to obtain the necessary insulation and connection. By placing the wires, properly insulated, in channels in the upper part of the conduit, on each side of the slot, they can be protected above and at the sides from dirt and wet; connection with the motor being made through the bottom of each channel; so that, with proper drainage, the insulation of the wires can be maintained; and by covering the motor connections with insulating material they can also be protected from electric loss through contact with snow or mud.

Elevated-Road Distribution.—On elevated roads the positive conductor can be connected with a central rail, both being properly insulated, and the track rails used for the return circuit, or any other convenient, economical method adopted; insulation and protection being comparatively easy, presenting no such difficulties as surface roads.

Thermo-Magnetic Motors.—The construction of these motors, which is still in the experimental stage, depends on the well-known principle that the heating of a mag-

THE DYNAMO AND MOTOR. 205

net reduces its magnetism; hence if a rotary, iron armature be mounted between the poles of a powerful electromagnet and its opposite sections, at points unequally distant from each pole, be alternately heated and cooled, a constantly varying magnetic force is developed, and rotation produced and maintained; the amount of force, speed of rotation, and permanency and economy of the apparatus being dependent on the construction. The capacity of the experimental motors of this class, constructed by Tesla, Edison, Menges, and others, is quite limited.

Such machines may also, like electric motors, be employed as electric generators, by rotating the armature in opposition to the magnetic force.

The economic value of a practical motor of this kind, by which the direct conversion of heat into power could be accomplished, would, for some purposes, be very great; since the present system of converting heat into power by the steam-engine, power into electricity by the dynamo, and electricity into power again by the motor, is wasteful and expensive, notwithstanding its many advantages: and if the elimination of the steam-engine and the dynamo could be accomplished, and the energy developed by the furnace utilized without loss by the substitution of a magnetic or electromagnetic engine for the steam-engine, a very desirable end would be attained.

Still it is not probable that such an apparatus could supersede the use of the dynamo and motor as a convenient, economical means for the distribution of power from a central station for running cars and many other purposes.

CHAPTER VIII.

ELECTROLYSIS.

WE have seen that the electric development which takes place in a battery cell is proportional to the chemical reaction, and, conversely, it is found that the chemical reaction developed by an electric current derived from the cell, or otherwise, is proportional to the electric development. In explaining polarization it was shown how water may be decomposed by the electric current. This decomposition was discovered by Carlisle and Nicholson in 1800, and it was subsequently ascertained that many other chemical compounds could be decomposed in a similar manner.

Nomenclature by Faraday.—Faraday, who made a very thorough investigation of this subject, gave to this process the name of *electrolysis*, a term derived from $\lambda \upsilon \omega$, to loosen, or separate, combined with $ηλεκτρον$, and he called substances capable of such decomposition *electrolytes;* hence the term *electrolytic* is applied to the cell in which the process is conducted, and sometimes also to the products of the decomposition, to distinguish them from the same substance obtained by other means; as an "electrolytic" metal.

The term *electrode*, which has already been defined, is also due to Faraday, and was first given to each of the wire terminals of the electric circuit connected with the electrolytic cell, though its use in other connections has since been found convenient, as already shown. He called the terminal by which the current enters the

ELECTROLYSIS.

cell the *anode*, and that by which it leaves, the *cathode;* the former term being derived from ανα οδος, ascending way, and the latter from κατα οδος, descending way.

He gave the name *ions* to the products of the decomposition, designating those which appear at the anode, or positive pole, as *anions* and those which appear at the cathode, or negative pole, as *cations*. Hence the anions are regarded as electronegative, and the cations as electropositive; each being attracted by the pole whose electric potential is supposed to be different from its own, and repelled by the one whose electric potential is supposed to be the same; electric energy overcoming chemical affinity. In the electrolysis of water, oxygen, appearing at the anode, is regarded as electronegative, and hydrogen, appearing at the cathode, is regarded as electropositive.

Theory of Grotthuss.—The transfer of these atoms, or " migration of the ions," in opposite directions, which is a salient fact, is supposed to occur in accordance with a theory proposed by Grotthuss in 1805, and subsequently modified by Clausius. In every liquid a mutual interchange of relationship is supposed to be constantly occurring among the molecules and atoms which compose them, producing motion in every conceivable direction, old groups being continually dissolved and similar new ones formed, and constancy of constitution thus maintained under continual change.

In Fig. 83 we have, in line 1, an ideal view of this heterogeneous movement; each little oval representing a molecule of water, the two hydrogen atoms, which compose the hydrogen molecule, being shown by the shaded part and the oxygen atom by the unshaded part, an infinite number of such chains making up the mass of the liquid. In line 2, these molecules, under the influence of an electric current from A to B, are supposed to be reduced to a symmetrical phase, in which the oxygen part of each is turned towards the anode and the hydrogen part towards the cathode. Now a new grouping is supposed to take place, the oxygen of each molecule moving to the left to recombine with the hydrogen of the adjoining left-hand molecule, and the hydrogen moving to the right to recombine with the

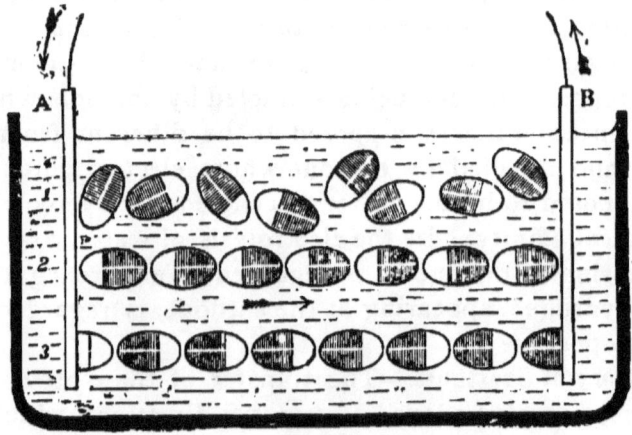

FIG. 83.

oxygen of the adjoining right-hand molecule; the new formation being represented by line 3, in which the oxygen part of the molecule at the left end of the chain and the hydrogen part of the molecule at the right end is each left without a mate, and being prevented by electric action from combining with each other, each is given off as gas. Thus while an interchange of atoms and molecules is taking place all along the infinite number of lines of electric action throughout the entire mass, the opposite ions make their appearance only at the electrodes. The decomposition of every other electrolyte is supposed to take place in a similar manner.

Electrolysis of Water. — Different compounds vary greatly in their relations to electrolysis, and the electrolysis of the same compound often shows great variation under different conditions. The feeblest current produces electrolysis in some cases, while in others the most powerful fails to produce it. Pure water, for instance, resists the strongest electrolytic action, while water slightly acidulated with sulphuric or chlorhydric acid is easily decomposed; the acid remaining apparently unchanged, while its presence reduces the electrolytic resistance of the water.

It has been suggested, in explanation of this, that there is a decomposition and recomposition of the acid, in this connection, in such a manner as to leave it unchanged; the decomposition of the water being indirect, through the agency of the acid; one or both gases being derived from the acid, which in turn receives from the water the same amount of one or both which it has surrendered. In case sulphuric acid (SO_4H_2) is used, it could furnish both; but in case chlorhydric acid (HCl) is used, it could furnish only hydrogen, while the hydrogen, taken from the water to replace this, would set free the proper combining proportion of oxygen. The theory given above shows how this may occur.

It is evident that whether the decomposition of the water is direct or indirect, the final result would be just the same; the two gases being evolved in the exact proportions in which they recombine to form water.

The high resistance of pure water to electrolysis does not absolutely prevent its decomposition. Gladstone and Tribe have effected it with zinc coated electrolytically with spongy copper or spongy platinum, also with iron or lead similarly coated with copper: but, in this case, the electrodes being intimately connected, the resistance is reduced to a minimum, while decomposition and the evolution of both gases with two platinum electrodes, separated, has not been found possible. But where the anode is an oxidizable metal, as copper, with which the oxygen, in the nascent state, can unite chemically, the decomposition may be effected. This also occurs when sodium or potassium is brought into contact with water, the oxygen uniting with the metal and hydrogen being given off.

Authorities differ in regard to the electrolysis of water under variation of pressure. It has been maintained that electrolysis is influenced by pressure much in the same manner as evaporation is thus influenced; that under a pressure of 300 atmospheres—about 4500 pounds to the square inch—even acidulated water cannot be decomposed, while in vacuo its decomposition may be effected by currents too weak to effect it under ordinary atmospheric pressure. But Bouvet claims to have effected it under a pressure of several hundred atmospheres, and to have found that the amount decomposed was independent of the pressure.

As water is the usual solvent in solutions, its electrolysis is usually inseparable from that of the substances held in solution, and becomes an important factor in the work required.

Conditions of Electrolysis.—The required conditions of electrolysis are that the substance must be a liquid, either naturally or by liquefaction, a conductor of electricity, and a compound, one of whose constituents is usually a metal. Ice, though of the same chemical constitution as water, and a conductor of electricity, cannot be electrolyzed, because it is a solid. All the oils, and nearly all melted fats and resins, being non-conductors, are not subject to electrolysis: carbon bisulphide, the liquid chlorides of carbon, and many other substances belong to the same class. Solutions

ELECTROLYSIS. 211

of the salts of copper, silver, gold, potassium, and sodium are among the substances most easily electrolyzed.

The metallic elements usually appear at the cathode and are regarded as electropositive, and the nonmetallic at the anode and are therefore regarded as electronegative. Hydrogen, which is considered a metal, appears, as we have seen, at the cathode.

But in the liberation of the same element from different compounds, it may be either electropositive or electronegative according to the positive or negative character of its associate elements; positive and negative expressing merely relative differences of potential under different conditions, and not absolute differences of physical constitution.

Temperature has a very important influence; rise of temperature increasing both the electric conductivity and the electrolytic action.

The time in which the action takes place is also of great importance; the results of rapid action generally differing considerably from those of slower action. Thus a simple metal, deposited at the cathode, may vary considerably in structure, or an alloy may differ in its composition, according as the process of deposition is slow or rapid.

Secondary Reaction.—Secondary reaction often occurs in electrolysis by which the liberated ions form new combinations with each other or with the electrodes themselves. We have had an instance of the former kind in the supposed decomposition and recomposition of acid in the electrolysis of acidulated water, and of the latter, in the union of the oxygen of water with potassium, sodium, or copper, used as the anode.

Such secondary reaction may occur at either electrode or at both, with marked characteristics peculiar to each. At the anode, the most common phenomena are corrosion of the anode, evolution of gas, and the adhesion of the ions to the anode, either as simples or new compounds; while, at the cathode, the ions, liberated either in a solid, liquid, or gaseous form, may either adhere to the cathode, be absorbed, dissolve, or escape.

Alloys of the metals may also be formed at the cathode by the deposition of one metal upon another, also amalgams with mercury.

Hence the permanent products of electrolysis may differ greatly from the elementary substances liberated, owing to the formation of new combinations during the process.

The specific gravity of the liquid at each electrode often changes also, usually becoming heavier at the anode and lighter at the cathode.

Electrolysis of Mixed Compounds.—In the electrolysis of mixed compounds, the different elements are usually liberated in the order of their electropositive affinities; the least electropositive cation first, since it has the weakest chemical affinity, and the stronger ones subsequently, in proportion to increase of current strength, or as reduction in the size of the electrodes increases the potential difference or E. M. F.

But, by making the quantity of each substance in the solution proportionate to its electropositive strength, several elements may be liberated simultaneously; and, by increasing the proportion, the stronger may, in some instances, be liberated in larger amount than the weaker.

By varying the proportions and other conditions in this manner, Favre was enabled to obtain from a mixture of the sulphates of cadmium, copper, and zinc, each metal separately, and also two, or all three, simultaneously; and found that the various results depended on the energy of the battery, the electrolytic resistance of the salts, and the relative time of electrolytic action; and hence he concludes that, by thus varying the conditions, the different metals may be separated successively from any mixture of metallic salts capable of electrolysis.

Relations of Electrolysis to Heat.—The evolution of heat is a necessary result of all electrochemical work, and is due both to chemical action and to electric resistance. When elements combine chemically, as in the battery cell, heat is generated, and when they are separated electrically, as in the electrolytic cell, heat is absorbed; and the amount thus generated or absorbed bears a certain definite proportion to the work accomplished and may be taken as its measure.

This heat is distinct from that generated by the electric resistance of the circuit, which varies in proportion to the amount of that resistance, and hence may be modified or controlled, while that due to chemical action is beyond control.

In the battery cell there is always, in connection with the chemical reunion by which heat and current are generated, a certain amount of electrolysis by which heat and current are absorbed; and in the electrolytic cell there is always, in connection with the chemical separation by which heat and current are absorbed, a certain amount of chemical reunion by which heat and current are generated; and in both cells there is also the generation of heat by electric resistance. Hence when heat is absorbed, in either cell, there must be corresponding electrolytic action, and, conversely, when such electrolytic action is developed there must be corresponding absorption of heat.

The electrochemical work required for the electrolysis of any compound must be equal to that required to develop the amount of heat which would be generated by its chemical recombination, plus that required to overcome the electric resistance of the circuit. Hence the heat developed by electrochemical action in the battery is the measure of the electric work accomplished by the current, minus that expended in overcoming the electric resistance of the circuit; otherwise the results would not be in accordance with the law of the conservation of energy.

Lowest Required Electromotive Force.—It has been shown that, in polarization, electrolytic action opposes electric generation; in like manner the action of the electrolytic cell opposes that of the battery, and when the opposing forces are equally balanced action in both must cease. Hence it is impossible to produce electrolysis with a battery whose E. M. F. is only just equal to that of the electrolytic cell.

If, for instance, the cell contain acidulated water, whose electrolytic reaction is $1.49\frac{1}{2}$ volts, its electrolysis, with platinum electrodes, would be impossible with a current from a single Daniell cell, whose E. M. F. is only about 1 volt; hence two such cells would be the least number by which it could be effected, or a single cell having a higher E. M. F. than $1.49\frac{1}{2}$ volts, as a Grove or a Bunsen. The minimum E. M. F. required in each case varies with the nature of the compound to be electrolyzed, but it must always be in excess of that of the electrolytic cell, unless re-enforced by secondary action in that cell.

When the anode is soluble and forms a new chemical combination with the liberated anion the minimum E. M. F. required for the battery is greatly reduced. This is the case when a copper anode is used in the electrolysis of acidulated water; the chemical reaction, producing combination of the oxygen and copper generates a current which re-enforces that of the battery, making the electrolysis of even pure water possible, as already shown. In this case the water may be decomposed by a single Daniell cell, or even one of less E. M. F.

It is claimed that polarization does not occur with an anode of the same metal as that deposited on the cathode, and hence that a current of the lowest E. M. F. will produce electrolysis under these conditions.

The opposing current set up in the electrolytic cell does not rise at once to its full E. M. F. Hence electrolysis may begin with a current of less E. M. F. than the required minimum, but cannot continue. This incipient electrolysis has been attributed by Helmholtz to the presence, in the solution, of such uncombined atoms as, according to Clausius, have become separated from their former associates, but have not yet formed new combinations; hence their segregation can be effected by a current of less E. M. F. than that required to separate atoms already combined.

Faraday's Laws.—The following laws were established experimentally by Faraday:

1. *The quantity of an ion liberated in a given time varies directly as the strength of the current.*

2. *The weights of the different ions liberated from a series of different solutions by the same current in the same time vary directly as their chemical equivalents.*

3. *Electrolysis is independent of the relative position of the electrolytic cell in the circuit.*

4. *The number and amount of chemical equivalents which enter into combination in the battery are equal to the number and amount liberated by electrolysis in the circuit.*

It is immaterial from what electric source the current is derived. Faraday produced electrolysis even with the slight current from an electrostatic machine; and Sir Humphrey Davy, in 1807, separated the metals potassium and sodium from their bases, for the first time, by the powerful current of a voltaic battery of 274 cells.

It is still an unsettled question whether an electric current can pass through a liquid without producing electrolysis. Observation seems to show that in some instances this may occur, and that in others the electrolytic effect is small in proportion to the conductivity.

Magnetic Effects.—Neither is it known to what extent magnetism influences electrolysis, as observation on this point has been very limited; but experiments by Remsen show certain marked peculiarities of manner in the deposition of copper, from its sulphate, under magnetic influence, which vary in proportion to the magnetic force, though the amount deposited remains unchanged.

Peculiar magnetic effects have also been observed by S. P. Thompson in the deposition of lead. In 1826 Nobili observed that the deposition of lead, from a solution of its acetate, upon a platinum anode, occurred in the form of rings which gave rise to very beautiful chromatic effects, and are known as *Nobili's rings*. Thompson has found that when the deposition is made in a magnetic field, it ceases to have the circular form, and assumes a form peculiar to the magnetic influence.

Chemical Equivalence.—Faraday's second law may be illustrated as follows: There is in every molecule of water 2 atoms of hydrogen and 1 atom of oxygen, but each oxygen atom weighs 16 times as much as each hydrogen atom, hence the *chemical equivalent* of oxygen is 8, that of hydrogen being 1 ; its volume being only half that of hydrogen, though its atomic weight is 16 times as great: that is, a given volume of oxygen, as a cubic foot, weighs 16 times as much as the same volume of hydrogen, but there are 2 cubic feet of hydrogen in a given volume of water for every cubic foot of oxygen;

and the liberation of these elements by electrolysis is therefore in this ratio.

Hydrogen being the lightest of all known substances, its chemical equivalent is taken as the standard of comparison for the chemical equivalents of all other substances. The chemical equivalent of copper, for instance, is $31\frac{1}{10}$, that being the weight of its atom as compared with that of 2 atoms of hydrogen; and the chemical equivalent of silver is 108, that being the weight of its atom as compared with 1 atom of hydrogen.

Now let the same electric current be passed, for the same time, through three vessels, one containing acidulated water, another some salt of copper, as its sulphate, and the third some salt of silver, as its nitrate; and, at the end of the time, let the products be weighed, and it will be found that for every gramme of hydrogen liberated there have been $31\frac{1}{10}$ grammes of copper liberated, and 108 grammes of silver.

Electrochemical Equivalence.—The weight of any substance liberated by a current of 1 ampere in 1 second is known as its *electrochemical equivalent*, and this is found to correspond practically with its chemical equivalent, in accordance with Faraday's law. Hence, if the chemical equivalent of any substance be multiplied by the electrochemical equivalent of hydrogen, the product is the electrochemical equivalent of that substance.

The electrochemical equivalent of hydrogen is found to be 0.000010352 of a gramme; multiplying the chemical equivalent of copper by this, we get $31.7 \times 0.000010352 = 0.0003281584$ of a gramme as the electrochemical equivalent of copper. In like manner the electrochemical equivalent of silver is found to be 0.001118016.

Effect of Current Reversal.—Faraday's third law must not be understood as applying to the relative positions of anode and cathode with reference to the direction of the current. If both are of the same substance and merely serve as conductors, as the platinum electrodes used in the electrolysis of water, their relative position is, of course, immaterial; but if they are of different materials, one or both of which is soluble, reversal of relations by change of current or otherwise changes the results. Such reversal, during the process, removes the ions already deposited on the electrodes. Hence an alternating current is not adapted to electrolysis.

Effect of Convection.—In a perfectly homogeneous solution the strength of the current is the same in every part, and hence the liberation of the ions is uniform; but the different parts of a solution are liable to a change of density during the process of electrolysis, producing differences in the liberation of the ions at different points on each electrode. This is especially the case when vertical electrodes are employed, with a metallic salt as the electrolyte. The specific gravity of the upper and lower parts of the solution, in proximity to each electrode, changes in consequence of difference of saturation; increase of saturation occurring at the anode with descent of the more highly saturated portion of the electrolyte, and decrease of saturation at the cathode with ascent of the less saturated portion, which has been deprived in part of its metal. This convection produces difference of resistance, causing the main direction of the current to be from the upper part of the anode to the lower part of the cathode; in consequence of which there is increased deposition of metal on the lower part of the cathode and a more rapid consumption of the upper part of the anode.

This is more especially the case when a strong current is employed; action being more uniform with a weak current. It is also more uniform with a horizontal position of the electrodes, also with solutions of a viscous character, in which this convection occurs more slowly.

Relative Conditions of Current and Electrolyte.—According to Quincke, electrolysis is proportional to the strength of the current per unit of sectional area of the electrolyte, varies with the E. M. F., is inversely proportional to the distance between the electrodes, and is independent of the cross-section and conductivity of the electrolyte, when the resistance of the rest of the circuit is small in comparison.

The conditions of electrolysis thus far considered are those in which a current from an external source passes through an electrolytic cell, but it may also be effected, to a limited extent, by currents generated by contact between the electrodes and electrolyte, as follows: 1. By dipping a metal into a solution consisting of a single liquid, as iron into a solution of copper sulphate or nitrate, which results in the deposition of a thin film of copper upon the iron; the coppering of iron wire being done in this way. 2. By employing two solutions of different specific gravities, the lighter one resting on the surface of the heavier, and using, as in the former case, a single electrode in contact with both solutions. The current generated between the two liquids produces deposition on that part of that metal which, under the conditions, becomes the cathode; the same result being produced by separating the solutions by a porous cell or partition, and bending the metal so that its opposite ends dip into each solution. 3. By immersing, in a single solution, two metals, in contact externally, or connected by a conductor; this produces a current from one metal to the other within the cell, causing electrolysis, the circuit being completed through the external connection. 4. By immersing two metals, connected externally, in two solutions separated by a porous cell or partition.

These various methods will be recognized as instances of the partial electrolysis, already referred to, which occurs in every battery; the cells being in fact various styles of battery cells.

Electroplating.—The electrolytic deposition of a metal upon another metal is termed *electroplating*, and is the principal means by which plating is now accomplished; its first introduction as an art being by Richard Elkington of England, in 1840, though it had been performed experimentally by Wollaston in 1801, and by Brugnatelli in 1805.

Various Details.—The principal metals used for this purpose are gold, silver, nickel, and copper, though other metals also are deposited in this way, as platinum and tin; also alloys, as brass, bronze, and german-silver. It is estimated that 125 tons of silver are used annually for electroplating in different parts of the world, 25 tons being thus used in Paris alone.

A vat of suitable size is provided, usually made of pine, and lined with lead or gutta percha; enamelled cast-iron is also used for this purpose; and, for small establishments, vessels of glass, china, or stoneware are used.

The current is furnished by a battery or other generator, usually a dynamo in large establishments, and must be always maintained in the same relative direction.

The solution consists of some salt of the metal to be deposited, which yields it pure in sufficient quantity, and in the most economical and efficient manner; and the solvent is strictly pure water, usually distilled rainwater.

The anode consists of one or more plates, usually of the same metal as that which is to be deposited, while the articles to be plated constitute the cathode; and both electrodes are suspended in the solution from copper bars which rest on copper strips, insulated from each other and connected respectively with the terminals of the generator, as shown in Fig. 84.

FIG. 84.

The Anodes.—The surface area of the nickel anode plates, used for nickel-plating, should equal or exceed that of the cathode surface, each cathode surface being exposed to an anode surface, and the depth of submersion of the anodes should be about two thirds the depth of the solution. For plating with gold and silver, less anode surface is required, and the exposure of both surfaces of the cathode is less important, as the solutions part with their metal more easily than the nickel solution, and the anodes are more soluble.

For nickel-plating plane, even surfaces, the distance

between the surface of anode and cathode should be about 3 inches, but for surfaces having prominences or cavities the distance should be increased to 5, 6, or even 10 inches, according to the amount of unevenness. For silver-plating, this distance should never be less than 4 inches. It is evident that while the distance, in either case, should not be such as to produce too great resistance, increase of distance tends to greater evenness of deposit on uneven surfaces, the ratio of difference in distance produced by such surfaces, as compared with the entire distance, being proportionally reduced by increase of distance, producing greater evenness of electrolytic action.

Hooks of nickel wire are used for the suspension of nickel anodes, but copper wire may be used if it does not come in contact with the solution; and these hooks should be sufficiently numerous to insure full conductivity. Silver anodes are suspended by iron wires or lead ribbons, and completely submerged, to prevent corrosion of the anode at the surface of the solution.

As the metal of the anode, when the same as that deposited, replaces that taken from the solution, its purity is a matter of great importance, affecting the color, brilliancy, and general quality of the work. But insoluble anodes of a different metal, or of some other substance, are sometimes used with advantage. Platinum anodes are especially recommended for nickel-plating, being indestructible. But their exclusive use is not desirable, as the electrolytic resistance is greater with insoluble anodes, requiring increased expenditure of electric energy to overcome it; besides, the metallic constancy of the bath is continually changing, being weakened by the abstraction of the metal, requiring repeated additions to maintain the requisite degree of saturation. Hence it is important that a certain proportion of the anodes, usually about one third, should be of nickel, both to maintain constancy and reduce resistance.

Carbon anodes may be used where platinum is considered too expensive. But even the best carbon anodes are liable to disintegration, and require to be renewed from time to time, which is a serious objection to their use.

The beautiful deposits of green, red, and pink gold, seen on watchcases and jewelry, are obtained by the use of silver anodes for the green, and copper anodes for the red and pink, the operation being finished with a gold anode of the same color as the deposit.

Plating Solutions.—The solution most generally used for nickel-plating consists of a double salt of nickel and ammonium, obtained by mixing in proper proportions, in distilled water, either nickel chloride with ammonium chloride, or nickel sulphate with ammonium sulphate. For certain purposes the character of the solution is modified by the addition of a little citric or chlorhydric acid.

Solutions for silver-plating are composed variously as follows: silver nitrate; silver potassic cyanide; silver chloride and potassic cyanide; chlorides of silver and of soda; cyanides of silver and of potassium; silver nitrate, potassium carbonate, and ferricyanide of calcined potassium.

The solution generally used for gold-plating consists of gold chloride combined with potassium cyanide; the chloride being obtained by dissolving the pure metal in a mixture of 2 parts chlorhydric and 1 part nitric acid, known as *aqua regia*.

Auxiliary Operations.—The preparation of articles for plating and their subsequent finishing are among the most important parts of the process, requiring a series of operations which cannot be described here in full.

The principal preparatory steps are termed *buffing, cleansing, pickling,* and *scouring.* The buffing consists in polishing the surfaces by means of revolving disks and brushes, and finely powdered substances, as fine sand, pumice, emery, lime, and crocus, before plating, and in polishing nickeled surfaces in a similar manner after plating

The cleansing, which is one of the first operations, is done by immersion of the article in hot potash or caustic soda; the pickling by its immersion in water acidulated with sulphuric acid; and the scouring by its immersion in a bath of nitric and sulphuric acids, to remove any remaining traces of oxide. Special baths are also used with different metals for scouring and other purposes.

Zinc is given a light coating of copper before nickel-plating, by immersion in a solution of copper acetate, combined with salts of soda and potassium; this being necessary to procure adhesion of the nickel; and iron is sometimes similarly coated by immersion in a solution of copper sulphate and sulphuric acid. Zinc may also be amalgamated as a preparation for nickel-plating; and for silver-plating, amalgamation is a prerequisite for all metals, the bath for this purpose consisting of a solution of mercuric binoxide in water acidulated with sulphuric acid.

Articles prepared in the above manner are suspended on hooks of suitable metal before immersion in the last preparatory bath, and must not be touched with the hand again before immersion in the plating bath, as the slightest contact of the bare hand, in nickel-plating especially, greases the surface sufficiently to produce an imperfect spot.

The time of immersion in each preparatory bath varies from a few seconds to 15 minutes, the longest time being required for the potash bath, and the operations being accompanied with frequent rinsings; and, after plating, the articles are rinsed in water and dried in hot sawdust before polishing or burnishing.

Required Electric Energy.—The electric energy of E. M. F. and current strength required varies with the metal to be deposited. For nickel-plating especially it should be vigorous at the beginning and weaker towards the close, the E. M. F. varying from 5 volts to 1 volt; and when furnished by a battery, 3 Bunsen cells in series, or their equivalent, may be used at the beginning, and 1 Smee, or its equivalent, at the close.

For silver-plating, an E. M. F. of not more than two or three volts is employed at the beginning, and a current strength of 50 amperes per square meter of cathode surface.

For gold-plating the E. M. F. must not exceed one volt, as the solution has very low resistance, and the current strength must not exceed 10 amperes per square meter of cathode surface.

Required Time of Immersion, and Thickness of Deposit. —The time of immersion in the plating solution varies with the metal to be deposited, with the metal to be plated, with the thickness of deposit required, and with the source of current employed.

For nickel-plating, the time with a dynamo current varies from 15 minutes to an hour, and with a battery current from 2 to 5 hours. The average deposit is about 2 grammes per square decimeter, which gives a thickness of about $\frac{1}{40}$ of a millimeter. Heavier plating is liable to peel, unless special precautions are used; but the hardness of nickel renders heavy plating unnecessary.

The time for silver-plating varies from 3 to 4 hours with the dynamo current, and from 8 to 12 hours with the battery current. The average deposit does not exceed 3 grammes per square decimeter; the average deposit on forks and similar-sized table-ware being from 80 to 100 grammes per dozen.

Gold is deposited with great rapidity; the practical difficulty of the process being greatly increased by the necessity of rendering it immediately successful. A few minutes' immersion is sufficient to insure a good surface, which is usually very thin; gold-plating covering more perfectly, and producing a better finish in proportion to thickness, than plating with other metals.

Agitation of the Solution.—In all kinds of electroplating frequent or constant agitation of the plating solution is important to insure the homogeneousness necessary for evenness of deposit; this agitation being sometimes maintained, in large establishments, by some special mechanical device. Special precautions are also necessary to insure good work on articles having deep cavities and sharp angles.

Electrotyping.—The deposition of copper by electrolysis for the production of copies of woodcuts and similar engraved surfaces, and also of type, is known as *electrotyping*, and is the process by which the metal plates called electrotypes, used for all the finer grades of book and map printing, are prepared for the press.

The details, which are comparatively simple, are briefly as follows: Impressions of the type or cuts are taken with a press on plates composed of beeswax and graphite, each plate having sufficient surface for several such impressions. The surface is then shaved smooth and even, and "built up" to a thickness of about $\frac{1}{18}$ of an inch by additions of the melted composition to all the blank spaces; after which it is brushed with finely powdered graphite. It is then covered with a thin coating of copper, precipitated upon it from a solution of copper sulphate by iron filings; and the plates, thus prepared, are suspended by copper hooks in a solution of equal parts by weight of copper sulphate and sulphuric acid in distilled water, at a distance of about 2 inches from anodes of pure copper, of equal surface.

With a dynamo-current, two hours' immersion gives a coating of the requisite thickness, but with a battery-current 12 to 14 hours is required. A single very large Smee cell furnishes a strong current of low E. M. F., about $\frac{60}{100}$ of a volt, which does not produce electrolytic resistance by decomposition of the water.

The copper coating is then stripped from the plates, being loosened by expansion with hot water, its reverse surface coated with solder, and melted type-metal poured over it so as to produce a plate about $\frac{1}{8}$ of an inch thick.

The different impressions are then cut from the large plates, straightened, planed smooth on the under surface, trimmed to symmetrical shape, and mounted, at the regular type height, on wood or metal bases.

Electric Refining of Metals.—The refining of various metals by electrolysis has become an important art. It consists in obtaining them in certain required states of purity from the crude smelted products, and extracting, by the same process, such percentage of the precious metals as they may contain.

Copper is one of the principal metals thus refined; the art having originated with Elkington, to whom patents for the electric refining of this metal were issued, in England, in 1866.

In a single copper refinery at Hamburg the average daily product, at a recent date, was 8 tons of refined copper, $2\frac{1}{2}$ tons of which were chemically pure; and the gold extracted in a single year equalled $1\frac{1}{4}$ tons.

The electricity in this refinery is generated with specially constructed dynamos, from which currents can be obtained in parallel or in series, the parallel current from the largest dynamo having an E. M. F. of about 4 volts and a strength of about 3000 amperes, and the series current an E. M. F. of about 8 volts and a strength of about 1500 amperes; the electric energy represented by the joint product of E. M. F. and current strength being the same in either case—12,000 watts.

Two series of 20 baths each are employed in connection with this dynamo; each bath having an anode surface of 30 square meters, making a total of 1200 square meters; the cathode surface being of equal amount, and the distance apart of the two surfaces from 2 to 4 inches. The anodes are thick plates of the crude metal, and the cathodes, plates of the chemically pure metal, about 1 millimeter thick. The solution consists of copper sulphate, and the deposit occurs in thick layers, which are easily removed from the cathodes. The copper thus deposited liberates its combining equivalent of sulphuric acid, which unites with the copper of the anodes, furnishing a supply of sulphate by which the constancy of the bath is maintained.

The precious metals are precipitated into the sediment; from which they are separated, and subsequently refined by a separate process.

Electric Reduction of Ores.—The separation of metals from their ores is another important application of electrolysis. As such separation cannot be made successfully from crude ores, they must first be reduced chemically to salts capable of being electrolyzed, and the success of the process and its economy consists largely in the proper preparation of the ore in this manner; different salts of the same metal, treated by different methods, yielding to electrolysis with different degrees of facility, and producing the metal in varying degrees of purity and in variable quantity with the same current. Hence the nature of the preliminary process is often the sole condition of success or failure.

Among the various ores reduced in this manner, the principal ones are those of zinc, lead, copper, silver, gold, aluminium, sodium, and magnesium.

Sir Humphrey Davy, as we have seen, obtained potassium and sodium from potash and soda, experimentally, in 1807; but the first practical application of electrolysis to the reduction of ores was made by Bunsen in 1854. He obtained aluminium, sodium, magnesium, barium, and other rare metals by this process in quantities comparatively large, operating on the chlorides of most of these metals.

His process consisted chiefly in submitting the fused chlorides to electrolysis in a glazed porcelain crucible, maintained at a red heat, and divided into two compartments by a porous earthenware partition reaching nearly to the bottom. By this method he electrolyzed aluminium and magnesium with a current of 15 to 20 volts, derived from a Bunsen battery, using electrodes of coke carbon, the metals going to the cathode and the chlorine to the anode.

H. E. S. C. Deville subsequently improved Bunsen's process for the reduction of aluminium. He mixed 2 parts by weight of aluminium chloride with 1 part common salt (sodium chloride) pulverized, fused the mixture at a temperature of 218° C., and electrolyzed it in a glazed porcelain crucible, maintained at a temperature of 183° C. He used a cylinder of charcoal for the anode, immersed in a portion of the fused mixture contained in a porous cell placed in the crucible; salt being added in this cell to fix the chloride and prevent its volatilization.

The cathode was a platinum plate, and two battery cells furnished sufficient electric energy, the resistance being very low. The deposit contained a percentage of common salt, which was subsequently removed by dissolving it in water; and the metal was further purified by successive fusions, and treatment with the double chloride of sodium and aluminium as a flux.

About 1885 C. E. Becquerel separated silver, copper, and lead from their ores by electrolysis; the silver ore being first reduced to a chloride and the copper and lead ores to sulphates.

Instead of a current derived from an external battery, he used the electrolytic cell as his battery, the liquid being a solution of the ore, while the electrodes were composed of zinc, iron, or lead for the anodes, and copper, tin, or carbon for the cathodes; grouping the cells as required for E. M. F. or current strength.

For copper he arranged them as gravity cells, a light solution of iron sulphate being superposed on a denser solution of copper sulphate, a cast-iron anode being placed in the iron solution, and a copper cathode in the copper solution, the deposit being made on the cathode.

The production of electric energy at economical rates by the dynamo has revolutionized these earlier methods of the electrolytic reduction of ores by battery currents which were too expensive to be practical; and their chief value now consists in indicating the nature of the required preliminary preparation. But even at the present comparatively low cost of electric energy, the electric process is not always the most practical or economical, and its application in any particular case must be determined by the attendant circumstances. Where water-power is cheap and abundant, and fuel expensive and scarce, its application is likely to be more practicable than where these conditions are reversed; power being easily convertible into electric energy, while the fuel required for the smelting process might make it more expensive than the electric process.

The chlorides are still the salts most generally employed, though the sulphates, nitrates, and acetates are preferable for some metals and for some processes. These salts are prepared from the ores by roasting, fusing, pulverizing, mixing with various substances, treating with acids, and other operations, according to the nature of the ore or the process; and are then reduced to the liquid condition for electrolysis, either by fusion or by solution in water, the nature of the ore or process, as before, determining the method required.

The Hall Process for Aluminium.—The process of C. M. Hall of Oberlin for the electric reduction of aluminium from its ores, patented in 1889, has been put into successful operation by the Pittsburg Reduction Company, resulting in the production of the metal, nearly pure, in large quantities and at a greatly reduced price.

It is substantially as follows:

A steel crucible lined with carbon contains a bath, lighter and more electropositive than aluminium, composed, by weight, of 234 parts calcium fluoride, or fluorspar; 421 parts of the double fluoride of aluminium and sodium, or cryolite; 845 parts aluminium fluoride, obtained by saturating hydrated alumina, Al_2HO_6, with hydrofluoric acid. The bath's chemical composition is represented approximately by the formula $Na_2Al_2F_8 + CaAl_2F_8$; to which is added 3 or 4 per cent of calcium chloride, $CaCl_2$, to prevent the abnormal increase of electric resistance, otherwise liable to occur from the formation of certain impurities.

The crucible being set in a furnace, the bath is fused at a red heat, and alumina in the form known as bauxite, an anhydrous oxide of aluminium, or the pure anhydrous oxide, Al_2O_3, artificially prepared, is dissolved in this fused bath and subjected to electrolysis with a dynamo current of 4 to 8 volts E. M. F., which is sufficient to electrolyze the alumina, but not the bath.

Carbon electrodes are employed, the anode being immersed in the bath and the carbon lining of the crucible forming the cathode. The aluminium goes to the cathode, sinking to the bottom on account of its greater specific gravity, where it can easily be drawn off; and the oxygen goes to the anode, where it unites with the carbon, forming carbonic acid, CO_2, which escapes as gas; the anode being thus consumed at the rate of about 1 pound of carbon to 1 pound of aluminium produced, and requiring frequent renewal. The presence of the calcium chloride prevents more rapid carbon consumption by suppressing the formation of carbonic oxide, CO, otherwise liable to occur, and which consumes double the amount of carbon, the oxygen uniting with the carbon in the proportion of 1 to 1 instead of 2 to 1, as shown.

The electrolysis proceeds continuously, aluminium being drawn off and alumina added in sufficient quantity to keep the bath saturated with it, though an excess is not injurious, as it merely sinks temporarily and is subsequently taken up. The bath also requires occasional additions of material to renew the loss due to volatilization and other causes; the calcium chloride volatilizing most rapidly, and its abnormal reduction being indicated by a fall of current due to increase of electric resistance.

CHAPTER IX.

ELECTRIC STORAGE.

The Leyden Jar and Condenser.—The storage of electric energy in the Leyden jar was one of the earliest discoveries in electric science. A full description of this instrument and its principles is given in the author's "Elements of Static Electricity," so that it is only necessary here to say, that it is simply a glass vessel, coated with metal on both surfaces to within a few inches of the top, which is left bare for insulation.

An electrostatic charge, positive or negative, given to either coating, usually the inner, by a static machine or spark coil, produces by induction a charge of opposite potential on the other coating, when connected with the earth or opposite pole of the machine or coil, and the two coatings remain in this electric condition till gradually restored to the same potential by the slow convection of the air; but an instantaneous discharge may be produced, attended with spark and snap, by making a connection between them by a conductor. And it is characteristic of this instrument, that while the charge is received gradually, occupying usually some seconds or minutes, the discharge, produced as above, is always instantaneous, and nearly complete.

The condenser, described in connection with the induction coil, is an instrument of the same character, receiving and surrendering its charge in a somewhat similar manner.

Grove's Gas Battery.—Polarization is another instance of electric storage, and observation of this phenomenon, and of the fact that the oxidation of the copper anode in an electrolytic cell produces similar storage, led to some experimental investigation of the subject by Gautherot and Ritter early in the present century. In 1842

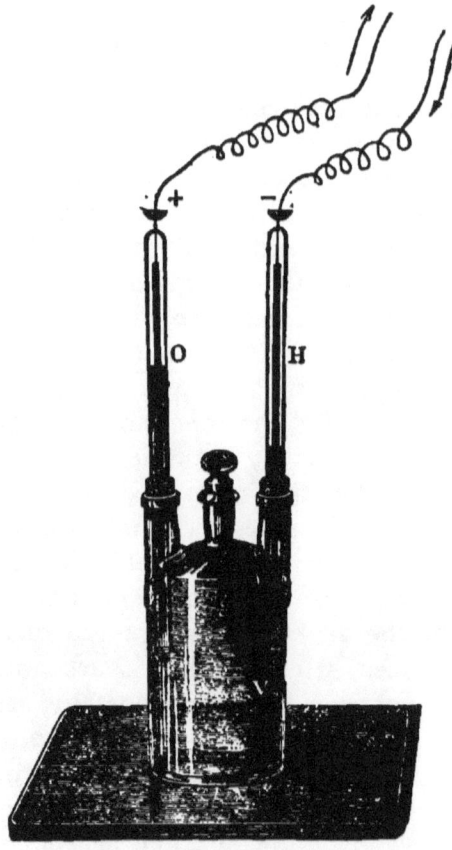

FIG. 85.

Grove constructed a gas battery on this principle, which is illustrated by Fig. 85.

A three-necked flask, V, contains acidulated water, into which are inserted two inverted tubes, containing respectively oxygen and hydrogen, designated by O and

H. Platinum wires, sealed into the upper ends of these tubes, are connected with platinum electrodes in contact with the gases above and water below, and the external circuit is completed through copper conductors whose terminals dip into mercury cups.

When the circuit is closed the gases recombine to form water, generating an electric current, which in the cell is from hydrogen to oxygen, and externally from oxygen to hydrogen, and whose E. M. F. is equal to that required to electrolyze water, $1.49\frac{1}{2}$ volts.

It is evident that the gases could either be evolved by a separate chemical process and admitted to the tubes previous to the latter being connected with the cell, or be evolved directly from the acidulated water of the cell by a battery or dynamo current. In the latter case the total amount of electric energy generated by the recomposition would equal that expended in the decomposition, and in the former the amount of electric energy obtained would be in the same proportion for the same amount of gas recomposed. In either case there is *storage of electric energy by chemical decomposition, which is recovered by chemical recomposition;* and this is the principle of chemical electric storage as developed in the various styles of the apparatus, known as the *storage battery, accumulator,* or *secondary cell.* The generation of electric energy must always follow the recombination, whether the elements are evolved in the gaseous form by insoluble electrodes like platinum, or in the solid form by combination with soluble electrodes, of which the oxide formed on a copper anode in the electrolysis of water is an instance.

Grove constructed similar batteries with other gases, and also with plates covered with metallic peroxides.

Wheatstone, Siemens, and Kirchoff made similar experiments, but Gaston Planté's discovery, in 1859, of the special adaptation of lead plates for this purpose, opened the way for the practical success of electric storage.

Planté's Secondary Cell.—Planté constructed a cell, using as electrodes two large sheets of lead rolled together and electrically insulated from each other with strips of gutta-percha, as shown in Fig. 86; the method

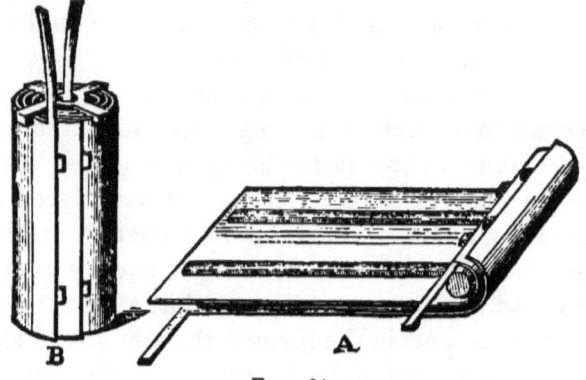

FIG. 86.

of rolling being shown at A, and the sheets, rolled and clamped, at B, projecting strips of lead being left attached to each for terminals. They were then immersed in water acidulated with ten per cent sulphuric acid, in a tall glass jar, and subjected to the action of a battery current supplied by two or more cells. A portion of the water being decomposed, the oxygen evolved at the anode combined with the lead, forming a dioxide, and the hydrogen was occluded on the cathode.

When the anode ceased to absorb oxygen, as indicated by the escape of the gas, the cell was disconnected from the battery, and discharged by making an external connection between the terminals of the electrodes, and then recharged with a reversed current.

This process was repeated during a period of several

months, the time of charging being continually increased from a few minutes at first to several hours subsequently, with long and increasing intervals of repose previous to each discharge and reversal; its object being to cover one of the plates with a thick coating of dioxide of lead, and the other with a coating of spongy lead of equal thickness.

Chemical Reaction.—The chemical reactions, as described by Gladstone and Tribe, are substantially as follows: The first charging produces only a thin film of the dioxide on the anode and of the hydrogen on the cathode; but the discharging changes the dioxide, PbO_2, which is insoluble in sulphuric acid, to monoxide, PbO, which is at once reduced to sulphate, $PbSO_4$, by the acid of the solution; the liberated oxygen atom uniting with the lead of the cathode and forming monoxide, which is also reduced to sulphate, as on the anode; the result of the discharge being a thin film of lead sulphate on each plate.

During the second charging the sulphate of the plate, now made the anode by reversal of current, is decomposed, the sulphuric acid, absorbed to form the sulphate during the repose and subsequent discharge, is restored to the solution, and the lead, thus liberated, combines with the oxygen liberated simultaneously from the water and forms the dioxide. The hydrogen, also liberated from the water, goes to the plate now made the cathode and decomposes its sulphate; restoring the sulphuric acid to the solution, and liberating the lead, which adheres to the plate as a spongy coating.

The respective results of each subsequent charging and discharging are the same as those just described; and as the spongy lead affords increased interior surface, the chemical reactions and formation of dioxide are proportionally increased.

But increased thickness of the dioxide produces increased resistance to chemical reaction; hence arises the necessity for the period of repose before discharging, during which the chemical reaction of the anode plate, by the strong affinity of the lead for oxygen, changes some of the dioxide to monoxide, which the acid immediately changes to sulphate, and thus the resistance is lessened.

There is also a resistance arising from an interior coating of sulphate, not reduced to dioxide by the charging, which remains in immediate contact with the plates and impedes the local action of repose, making longer periods of repose necessary as the coatings increase in thickness.

Hence the electric formation of the plates consists of three distinct processes, the charging for the formation of dioxide on the one and spongy lead on the other, the repose for local action, and the discharging for the production of sulphate on both; the plates when completed consisting respectively of lead dioxide and spongy lead adhering to interior supports of sheet lead; and subsequent charging, for practical use, is always in the same direction, alternation being discontinued.

The charge may be given either by a battery or a dynamo, usually the latter; the chemical reactions, when the cell is in practical use, being just the same as during the preparatory process; the electric effect being the absorption of electric energy by the conversion of sulphate into dioxide during the charging, which is recovered by the conversion of dioxide into sulphate during the discharge.

It has been observed that often when a partially discharged cell is given a short period of repose, the subsequent discharge shows increased electric energy. This is accounted for on the hypothesis that when the discharge is rapid some of the sulphate, formed on the anode from the dioxide, is reconverted into dioxide by the excess of oxygen developed, producing a proportional reduction of potential difference between the plates; but that during the short repose this dioxide is again reduced to sulphate and the potential difference restored.

The maximum E. M. F. of the Planté cell is about 2.54 volts. By means of a commutator of special construction, Planté could instantly join a battery of 20 such cells either in series or in parallel. He used the parallel connection for charging, which he accomplished with 2 Bunsen cells, the resistance, with this connection, being very low, and the series connection for discharging, by which he obtained a maximum current equal to that of 30 large Bunsen cells.

The duration of the discharge depends on the resistance of the external circuit, varying from a few minutes to several hours according to the amount of current required; and it ceases when the dioxide is all changed to sulphate, but should be terminated sooner to prevent injury to the plates from the excessive formation of sulphate.

The Faure Cell.—The tedious, expensive process requisite for the electric formation of the Planté plates led to the construction by Camille A. Faure, about 1880, of plates prepared by covering sheet lead with a paste made of red lead and sulphuric acid; the coating being confined to the surface by a covering of paper and by felt placed between the plates, which also served the purpose of insulation.

Thus prepared and rolled together, they were placed in a glass jar, in water acidulated with sulphuric acid, and the coating subjected to electrolysis with alternation of current, by which the red lead, known also as *minium*, Pb_3O_4, was changed in a few days to lead dioxide and spongy lead, on each plate respectively, and the cell was ready for practical use.

Chemical Reaction.—The chemical reaction, according to Gladstone and Tribe, is substantially as follows: On the immersion of the plates in the acidulated water, there is, at first, a purely chemical reaction, by which the minium on both plates is gradually changed, from the surface inwards, into a mixture of the dioxide and sulphate of lead, with evolution of water, thus, $Pb_3O_4 + 2H_2SO_4 = PbO_2 + 2PbSO_4 + 2H_2O$. But oxygen and hydrogen being liberated by the electric current, the oxygen goes to the anode and changes the sulphate into dioxide and sulphuric acid, thus, $2PbSO_4 + 2H_2O + O_2 = 2PbO_2 + 2H_2SO_4$. The hydrogen goes to the cathode, changing the dioxide to spongy lead, with evolution of water, thus, $PbO_2 + H_4 = Pb + 2H_2O$; and changing the sulphate to spongy lead and sulphuric acid thus, $2PbSO_4 + H_4 = 2Pb + 2H_2SO_4$; reversal of current being necessary to electrolyze the heavy coatings completely.

The chemical reaction of the discharge is the formation of lead sulphate on both plates, and that of subsequent charging the reconversion of this substance to dioxide and spongy lead, as before.

Defects of the Faure Cell.—While the Faure cell could be produced much more economically than the Planté, and was equal to it in electric energy, it had many serious defects which proved fatal to its practical success. The felt, preventing the free circulation of the fluid, seriously impeded electrolysis; it soon became corroded by the acid and partly removed in patches, and ceased to insulate. The coating failed to adhere properly, sloughing off and falling to the bottom. Hence in a short time the cell became worthless. But its invention demonstrated the possibility of practical success by some similar method of construction, to ascertain which the investigations of Brush, Swan, Sellon, Volckmar, and others were immediately directed.

Improved Faure Cell.—The result of these investigations was the production, about 1886, of an improved cell, the principal feature of which is the improved style of plate illustrated by Fig. 87, which consists of a lead grid, shown at *A*, having its openings wider at the sur-

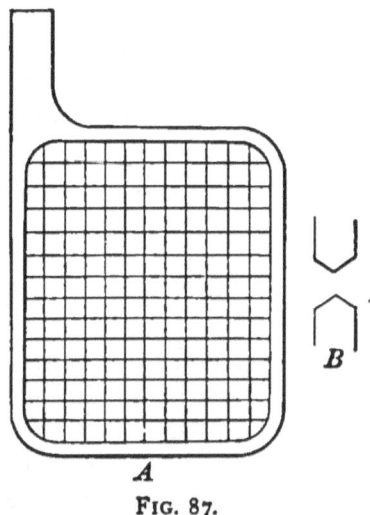

FIG. 87.

faces than in the interior, as shown by the enlarged section at *B*. These openings are filled with a paste made of minium and sulphuric acid for the positive plates, and of litharge and sulphuric acid for the negative; litharge being a red lead monoxide, PbO, more easily reduced than minium, for which reason it is preferred for the negatives, to facilitate the reduction of the paste to spongy lead, which is more difficult than its reduction to lead dioxide on the positive plates.

The advantages of this style of plate are that it gives a firm support to the paste, the plugs being held in the grids by the form given them by the openings, which obviates the necessity for the intervening felt and paper, allowing free circulation of the fluid and more perfect electrolysis.

The cells are made of different sizes, stationary and portable; the stationary cells having glass vessels, and the portable, hard-rubber vessels. The 23-plate stationary cell, shown in Fig. 88, has 11 positive plates and 12 negative; each set attached to a lead cross-bar above and at the center, by which the plates are held at a fixed distance apart; the two sets interlocking, so that positive and negative plates alternate and are insulated from each other by two rows of hard-rubber forks. Each plate is $\frac{1}{8}$ of an inch thick, and the space between adjacent, positive and negative plates, $\frac{5}{32}$ of an inch wide; and the two outside, negative surfaces being inactive, each set has 22 interior, active surfaces.

A thick plate of glass, under the central cross-bar and plate connections on each side, supports the plates, so as to leave a space underneath for the free circulation of the fluid; each set being supported, on its opposite side, by plate projections which rest on an insulating hard-rubber strip above each cross-bar as shown; two stout rubber bands holding these supporting plates and the lower parts of the lead plates in position. A lead bar, projecting from the cross-bar of each set, can be bent into any convenient position for making connection with adjoining cells.

These plates are immersed in water acidulated with

ELECTRIC STORAGE. 243

36% sulphuric acid, contained in a glass jar 10½ inches long, 8¼ inches wide, and 9⅞ inches high; the entire weight of jar and contents being 50 lbs.

FIG. 88.

The glass jar has the advantage of allowing inspection of the interior without disturbing the contents, by which the condition of the plates may be observed, and short-circuiting from the buckling of plates or the lodging of loose paste plugs between them remedied; but its comparative frailty and weight are objections to its use for the portable cells required on cars and elsewhere. Hence a portable cell of the same capacity and number of plates is constructed with a covered, hard-rubber jar, made shorter below than above so as to furnish supporting ledges for the plates at the opposite ends, which take the place of the glass supporting plates employed in the stationary cell. The weight of this cell is 40 lbs., its height about the same as that of the stationary cell, and its other dimensions about one fourth less.

The 15-plate stationary cell has 7 positive plates, each $\frac{6}{32}$ of an inch thick, and 8 negatives, each $\frac{6}{32}$ of an inch thick, contained in a glass jar $10\frac{3}{4}$ inches long, $12\frac{1}{4}$ inches wide, and $13\frac{3}{4}$ inches high; the entire weight being 130 lbs., and the storage capacity 300 ampere-hours, double that of the 23-plate cell. The supporting plates are of hard rubber, with openings for inspection, and are each held in position by two metal rods which pass through loops in the positives at one end and in the negatives at the other, binding the plates of each set together below and furnishing electric connection between them.

Electric Preparation of the Plates.—Each set of plates, positive and negative, is electrolyzed separately before they are combined in the cell intended for practical use; special sets of temporary plates, or dummies, of each kind being used for this purpose, which makes the old process by reversal of current unnecessary. The negatives, although thinner than the positives, require six days for the reduction of the litharge to spongy lead, while the minium of the positives is reduced to dioxide in 24 hours.

Electric Energy of Improved Cell.—The E. M. F. of this

cell is about 2 volts, and its internal resistance .001 to .005 of an ohm. Its current, as in the Planté, depends on the external resistance; 30 amperes for 10 hours being considered an economical working rate for the large 15-plate cell. If less current is required the time of discharge becomes proportionally longer; and a current of 300 amperes may be obtained for an hour, but such rapid discharge is injurious to the cell.

Effects of Charge and Discharge on the Plates.—As both the charge and discharge result in different forms of chemical reaction, it is obvious that ample time should be allowed for this reaction to produce the required chemical changes; the normal rate under varying conditions being ascertained better by practical experience than by arbitrary rule.

Charging is always accompanied by the evolution of gas, which, as has been shown, is chiefly absorbed, while a certain percentage escapes; hence if the rate of charging is excessive there is an abnormal escape of gas and useless consumption of current: there is also an abnormal development of heat, which may result in destruction of the plates.

As there can be no further absorption of gas when the chemical reaction of charging is completed, its abnormal escape with a normal current indicates an overcharge, resulting as before in waste of current, but not in injury to the plates.

But as the chemical reaction of the discharge results in the absorption of sulphuric acid and the formation of lead sulphate, a hard unyielding substance, in considerable quantity on both plates, and in excess of the material which it replaces, due to the absorption of acid, it is evident that if the action is too rapid, the plugs on the surfaces most exposed to chemical and electric action will become sulphated to a greater degree than on the opposite surfaces, producing unequal expansion, with warping, or buckling of the plates, as a result; the same result also occurring from an excessive formation of sulphate if the discharge is continued too long.

E. M. F. of Discharge.—The E. M. F., during the first half-hour of discharge, is about 2.25 volts, being slightly increased by the supplementary reaction of a small amount of gas which adheres to the plates after charging; it then drops to about 2.4 volts, remaining nearly constant, with a slight decline to 2 volts or less, till the dioxide is mostly reduced to sulphate, when it begins to decline rapidly; which indicates that the discharge should cease to prevent injury to the plates.

Conductivity and Buckling.—It is important that there should be a sufficient supply of sulphuric acid present to maintain the requisite conductivity during the discharge, when it is rapidly absorbed to form the sulphate; otherwise the fluid will soon be deprived of its normal quantity and the resistance abnormally increased. An excess of the acid is also injurious, causing the sulphate to form too rapidly, with buckling of the plates as a result. Such excess is liable to occur in the lower part of the cell, where the acid, from its greater specific gravity, accumulates, causing a corresponding reduction in the upper part. This increases the conductivity and chemical and electric action below, with corresponding decrease above. Hence buckling usually increases downward.

Buckling, if not excessive, and if in the same direction on all the plates, does not interfere with the action of the cell. But it always tends to loosen the plugs, so that they are liable to drop out and fall into the space between the plates, producing a short circuit. There is also liability to short-circuiting by contact between positive and negative plates, if the buckling is in opposite directions.

Weight of Cells.—When a storage battery is used on a car to furnish light or motive power, reduction of weight becomes highly important, as a large number of cells are usually required, and their aggregate weight will often equal 3500 pounds.

Composition of Grids.—Various metals and alloys have been tried for grids, but lead still has the preference, and is in general use. The addition of a small percentage of antimony, as a flux, aids in producing more perfect castings; lead alone failing to flow into the narrow spaces in the molds with the requisite facility. The further addition of a very small percentage of mercury to increase the durability, has also been tried, but its use has proved detrimental; and the antimony, though advantageous, as a flux, is not so durable as the lead.

The Julien Cell.—Various improvements of the Faure cell have been attempted, the chief objects of which have been to obtain greater durability, reduced weight, and to prevent the buckling of the plates. Prominent among these is the cell of Edward Julien of Belgium, whose general construction is similar to that of the improved Faure cell. Its chief claim is a special composition of superior durability for the grids, which, so far as can be ascertained, is 94.5 per cent lead, 4.2 per cent antimony, and 1.3 per cent mercury.

The Pumpelly Cell.—A cell has recently been invented by J. K. Pumpelly of Chicago, the principal features of which are a horizontal position of the plates, supporting material between them, copper electrodes centrally located on opposite sides of the cell and reaching to its bottom, and a containing vessel of light durable material. The construction, in other respects, is similar to that of the Faure cell, and the same materials are employed for the grids, paste, and fluid. A slight burr at the narrow part of the grid openings holds the paste more securely, 12 per cent of antimony enters into the composition of the grids, and 20 per cent of sulphuric acid into that of the fluid.

Fig. 89 shows the construction. The positive and negative plates alternate in position, the top and bottom plates being negative, and they are supported and insulated by cellulose made from wood, said to be a good insulator, an excellent absorbent, and indestructible in sulphuric acid. Each plate has, at the centre of one of its edges, an opening about an inch square, and, at the same point on the opposite edge, a round, vertical, tubular projection about an inch high and half an inch in diameter, on the under side of which is a small socket fitted to the upper end of a similar projection from the alternate plate underneath.

FIG. 89.

When the plates are built up in the cell, with the cellulose between them, each set has the projections all on the same side and the openings on the opposite side; the projections of each alternating with the openings of the other on the same side; so that each projection from a negative passes up through an opening in a positive, with ample space for insulation, and helps to support the next negative above; and each projection from a positive passes similarly to the next positive through an opening in the intervening negative. These projections form a continuous tube on each side, from top to bottom, in which are placed the copper electrodes, and melted lead is poured in around them, giving perfect metallic contact, and holding each set of plates firmly in position.

The plates, thus built up, are immersed in the fluid in a hard-rubber vessel, rest on wooden blocks, and are charged, without reversal of current, in the cell designed for use. The E. M. F. is about 2 volts.

Durability of Storage Cells.—Manufacturers usually guarantee for the positive plates a durability of one year in constant practical use, with a normal current. The negatives are far more durable, not being subject to oxidation; and, unless injured by buckling, last for an indefinitely long period.

Storage Capacity.—The storage capacity of the 15-L. Faure cell, or the 300-ampere Pumpelly cell, is about 1,080,000 coulombs, equal to 30 ampere-hours. Hence such a cell may be discharged in 1 hour with a 300-ampere current, or in 10 hours with a normal, 30-ampere current; the time in seconds or hours, for a normal discharge, being estimated at $\frac{1}{10}$ of the storage capacity in coulombs or ampere-hours.

Relative Time of Charging and Discharging.—The time required for charging a cell is estimated at 18 to 20 per cent more than that required for discharging it with the same current strength; that being the usual percentage of loss of energy between the charge and discharge. Hence if a 300-ampere cell is discharged in 10 hours with a 30-ampere current, 12 hours would be required to charge it with a 30-ampere current, or 36 hours with a 10-ampere current. The current strength required for charging is estimated at 5 amperes per square foot of positive plate surface.

The preparatory charging of the Pumpelly cell at the factory occupies only the same time as each subsequent charging in actual use. Hence only about 60%

of the litharge on the negatives is reduced, at first, to spongy lead; the remainder being gradually reduced by use; which probably accounts in part for the fact observed, that the cell increases in energy during the first six months of use.

The Hydrogen Alloy Theory.—A new theory of the electrochemical action of accumulators has been proposed by Dr. Paul Schoop, based on the following facts: It has long been known that certain metals, as platinum, palladium, mercury, and iron, combine, under certain conditions, with hydrogen; and on the theory that hydrogen is a metal, these combinations are regarded as alloys.

It is also well known that when platinum sponge, charged with hydrogen, is exposed to the air it becomes rapidly heated to redness by the absorption of oxygen; also that the charged cathode plate of an accumulator, when similarly exposed, has its temperature raised, from the same cause, often to the melting point. Hence Dr. Schoop assumes that the spongy lead of the cathode, like the platinum sponge, absorbs the hydrogen liberated from the solution by the electric current during the charging; the hydrogen combining with the lead and forming an alloy, and the liberated oxygen combining with the material in the anode and forming the lead dioxide: and that during the discharge, oxygen liberated by the current from the dioxide combines with the hydrogen of the cathode, reducing the alloy to spongy lead and restoring water to the solution; leaving the material in the anode with the same proportion of oxygen as before charging.

This theory is simple, but defective in failing to account for the formation of the lead sulphate, and its varying proportions during the charging and discharging. It is not easy to see how hydrogen can unite with spongy lead incrusted with sulphate; so that unless the formation of sulphate, under normal conditions of the cell, be ignored, or a cell produced from which its formation shall be eliminated, the correctness of this theory must be considered questionable.

CHAPTER X.

THE RELATIONS OF ELECTRICITY TO HEAT.

The mutual relations of heat and electricity are among the most important in electric science, whether considered with reference to the generation of electricity, its transmission, its measurement, or its numerous forms of practical application. There can be no expenditure of electric energy without the simultaneous development of heat; and it may also be assumed, though not so manifestly proved, that there can be no expenditure of heat energy without the simultaneous development of electricity.

Heat Developed by Electric Transmission.—According to the best evidence we have, electricity and heat are different kinds of molecular motion, and the transmission of either is simply the extension of this motion through a material substance connected with the generator, known as the conductor. When electricity is thus transmitted, its transmission is always attended by the evolution of heat, which must be considered a legitimate part of the work done, whether useful or otherwise, and not a mere adjunct.

This heat is found to be always in direct proportion to the electric resistance encountered; hence if the useful work to be done is the production of heat, or its concomitant, light, the resistance is increased at the point where the heat or light is required: but if other useful work is to be accomplished, the heat is suppressed by lessening the electric resistance, as required. Thus the ratio of heat work to other work can be made to vary by varying the resistance.

The analogy to this is found in the friction attendant on mechanical action, which may produce heat for a useful purpose, or be suppressed by the use of a lubricant when the mechanical energy is to be otherwise expended.

Joule's Law.—To determine accurately the relations between the electric current and the heat developed by it, Joule, who made this branch of electric science a specialty, passed a battery current through a fine wire coil inclosed in a vessel of alcohol, in which was also inserted a thermometer.

The resistance and current strength being known, were compared with the temperature to which the liquid was raised in a given time, and by this means were established the facts embodied in the following law:

The heat developed in a conductor by an electric current passing through it varies as the CONDUCTOR'S RESISTANCE, *the* SQUARE OF THE CURRENT'S STRENGTH, *and the* TIME THE CURRENT LASTS.

Representing the heat by H, the current by C, the resistance by R, and the time by t, we get $H = C^2Rt$ as the algebraic expression of this law, by which the heat developed in any electric circuit can be ascertained.

Joule's Equivalent.—Joule found that the amount of heat required to raise the temperature of 1 gramme of water 1° C. is equivalent, in work, to 42,000,000 ergs in C. G. S. measure; and this is known as *Joule's equivalent.* When the heat is produced by an electric current, the formula given above must be multiplied by 0.238 to reduce the electric C. G. S. units to heat units; that being the ratio, expressed decimally, of 10,000,000, the C. G. S.

value of the electric units represented by C^2Rt, to 42,000,000 (10,000,000 ÷ 42,000,000 = 0.238), and the formula is then $H = C^2Rt \times 0.238$.

Heat Developed by Electrochemical Action.—The experiments of Favre on the electrochemical development of heat fully establish the correctness of the principle, that the evolution of heat by electric action is in the inverse ratio of other work accomplished by the same action; and that the heat developed in the battery circuit is the exact equivalent of the chemical energy expended in the cells, as first verified approximately by Joule.

In these experiments he used a mercurial calorimeter, so constructed that the mercury should surround the apparatus in which the heat was to be generated, and by its expansion register the amount of heat developed. Placing in this instrument a vessel containing zinc and sulphuric acid, he found that the simple chemical consumption of 33 grammes of zinc produced 18,682 units of heat. He then replaced this vessel by a Smee battery cell, and noted the electrochemical consumption of the same amount of zinc; varying the experiments by using connecting wires of different resistance, and also by comparing the evolution of heat when the cell was placed in the instrument and the connecting coil was outside, with its evolution when the coil was placed in the instrument and the cell was outside. The results varied but slightly from that of the first experiment, the consumption of 33 grammes of zinc producing 18,674 units of heat. The first experiment showed the amount of heat developed by a given amount of chemical action, measured by the consumption of the zinc; and the second proved that practically the same amount of heat was developed in the battery circuit by this amount of chemical action in the cell.

To show the mutual relations between electric heat and other electric work, a battery of 5 Smee cells, joined in series, was placed in the calorimeter, and connected with a small electromagnetic engine; and the evolution of heat during the consumption of 33 grammes of zinc noted in three different experiments, as follows: 1. With the engine at rest the heat evolved was 18,667 units. 2. With the engine running, but doing no work except to overcome its own friction and inertia, the heat evolved was 18,657 units. 3. When the engine by raising a weight did 12,874,000,000 ergs of work, the heat evolved was 18,374 units. Dividing the number which represents the work in the last experiment by Joule's equivalent (42×10^6) gives 306 heat units, and $18,374 + 306 = 18,680$. Hence, with proper allowance for unavoidable discrepancies, we find that in the last three experiments, as in the first two, the heat evolved was equivalent to the chemical energy expended; while the last experiment proved that the evolution of heat is in the inverse ratio of other work; the heat which disappeared being reproduced as work; a result conformable to the doctrine of the conservation of energy.

Electro-Thermal Capacity of Conductors.—Since the heat developed in a conductor by an electric current varies as the resistance, and the resistance varies with the nature of the material, and also directly as the length and inversely as the cross-section of the conductor, it follows that *material, mass,* and *ratio between length and cross-section* must each be considered in estimating the conductor's *electro-thermal capacity.*

In conductors of equal length and cross-section but different conductivity, the heat developed in each by the transmission of currents of equal strength varies inversely as the conductivity, or, which is the same, directly as the resistance due to difference of material.

Thus german-silver having about 13 times the electric resistance of copper, the heat developed in a german silver wire would be about 13 times that developed in a copper wire of the same dimensions, carrying a current of equal strength.

But in conductors of the same material and mass, the resistance, and hence the heat development, varies *directly* as the ratio of length to cross-section, and *inversely* as the ratio of cross-section to length. Suppose 100 units of heat to be developed by the passage of a current through a wire 10 feet long, then only 10 units would be developed by the same current in a section of the same wire 1 foot long; hence if the wire be regarded as made up of 10 sections arranged in series, 10 units is the amount developed in each section successively. Now suppose a current of the same strength passed through a wire of the same material and mass, 1 foot long; the cross-section of this wire would evidently be 10 times as large as that of the other wire, consequently the resistance and heat development would be only $\frac{1}{10}$, that is 10 units; the effect being the same as if the current passed through the 10 sections in parallel. But as the 10 units are equally distributed through the mass, only 1 unit of heat is developed in each section; that is, $\frac{1}{10}$ the amount developed in each section, or same mass, of the long wire, or series connection.

Suppose a current of the same strength passed through a wire of the same material and mass, 100 feet long; the length being 10 times that of the original wire, the cross-section would evidently be only $\frac{1}{10}$; hence the resistance and heat development would be 10 times as great, equal to 1000 units, or 10 units to each foot. But these 10 units being developed in $\frac{1}{10}$ of the original mass per foot would raise the temperature to 10 times the original temperature per foot, or 100 units. Now since the cross-sections of wires vary as the squares of their diameters, and the heat development varies inversely as the cross-section, $\frac{1}{10}$ the cross-section producing 10 times the heat, it is evident that *the rise of temperature in a conductor, or heat development per unit of mass, varies inversely as the fourth power of the conductor's diameter.*

The *heat development per unit of mass,* as illustrated by the last example, deserves special notice. The number of heat units developed in a foot of the ten-foot wire was found to be just the same as in a foot of the hundred-foot wire, 10 units in each, though the rise of temperature in the last was 10 times as great, being inversely proportional to the reduction of mass. Hence each wire, if immersed in an equal mass of the same liquid, to which its 10 heat units should be imparted, would produce the same rise of temperature, as indicated by a thermometer; for though the section of small wire becomes 10 times as hot as that of the large wire, it has only $\frac{1}{10}$ of the mass, and hence only the same heating power.

These principles have a highly important useful application, especially in electric lighting, which will be separately considered in a future chapter; but there are several minor applications, some of the more important of which may be noticed here.

Electric Blasting.—The explosion of a blast can be safely and expeditiously effected at any required distance, by inclosing a fine wire of high resistance, usually platinum, in the fuse, and connecting it with a battery circuit of low resistance, conveying a current of the required strength. When the circuit is closed, the current, which produces scarcely a perceptible change of temperature in the main conductor, instantly raises the platinum wire to a white heat, producing the explosion.

In this way blasts under water are fired, and mines and torpedoes exploded. The explosion of the great blast under the ledge of rock in Hellgate, New York harbor, by the touch of a child's finger closing the circuit, is a noted instance of this.

Electric Cautery.—In surgery a fine platinum wire, heated to incandescence by an electric current, is preferred to the knife for certain purposes; the operation, which is known as *electric cautery*, being more safely and expeditiously performed in this way; as in amputation of the tongue for cancer, the removal of an excrescence, or of superfluous hair from a lady's face.

Electric Fuses.—As conductors carrying strong currents are liable, from accidental causes, to become overheated and ignite inflammable matter in close proximity, a short section of a soft compound metal of high resistance, technically known as a *fuse*, is introduced into the circuit at any convenient point. The cross-section of this fuse is so adjusted to the normal strength of the current carried, that if, from any abnormal increase, the temperature approaches an unsafe degree, the fuse melts and opens the circuit.

The metals forming the compound are usually lead, tin, bismuth, and antimony, combined in different proportions according to the melting temperature, and other properties required. Fuses are usually from $\frac{1}{2}$ to $\frac{3}{4}$ of an inch long, and from $\frac{1}{40}$ to $\frac{1}{4}$ of an inch in diameter, and adjusted to carry currents of from $\frac{1}{2}$ an ampere to 200 amperes, without fusion; the melting temperature being made, by adjustment of cross-section, about 20 per cent above the carrying temperature in the large fuses, and about 50% above in the small ones, when inclosed. The reason of this is found in the nature of the composition required for each; bismuth and lead, which melt at a comparatively low temperature, entering largely into the composition of the small fuse to give it the requisite tenacity, while tin and antimony, which have a higher melting temperature, but are more brittle and less expensive than bismuth, predominate in the large fuse, in which there is less risk of fracture, and in which economy of material is less of an object. In the open air the melting temperature of the large fuses is about 5% higher than when inclosed, and that of the small ones about 8% higher.

As the conductivity of metallic conductors decreases with rise of temperature, and as the radiation of heat increases with increase of cross-section, both these points must also be considered; so that the proper construction of fuses, including material, cross-section, carrying capacity, and melting temperature, adapted to varying conditions, is a difficult scientific problem, and one of great practical importance. If a fuse melts too easily it becomes a source of annoyance from frequent, unnecessary interruption of current, while if its temperature of fusion is too high it fails to afford protection against fire.

Fuses are connected by binding-screws to insulating blocks, to which the terminals of the conductor are also similarly attached, and hence are easily replaced at a nominal expense when melted; several fuses, connected with different circuits, or different branches of a circuit, being often arranged in the same block.

Thermo-Electric Generation.—Before entering fully upon the consideration of thermo-electric generation, it is important to present certain general principles of electric generation which have a special bearing on this branch of our subject.

An examination of the various kinds of apparatus by which electricity is generated shows that the construction, in every case, involves the following conditions: 1. *A complete insulated electric circuit composed of heterogeneous materials.* 2. *Molecular excitation at one or more points in this circuit.* And it may be safely assumed that in all cases where these conditions are fulfilled, either by natural or artificial means, electricity is generated, even though such generation may not be apparent.

These conditions are a legitimate consequence of the law of the conservation of energy as applied to electricity considered as a mode of molecular motion. For if the circuit were not complete, molecular motion, excited at any point, must soon cease; for the continuous storing of energy in one place implies its removal from another place, and this cannot continue indefinitely, nor for any considerable time, without a connection between the two places by which the transferred energy can return to the place of its origin. The same would be true if the circuit were complete but perfectly homogeneous throughout, both as to material and resistance, for molecular motion would then be transmitted equally in opposite directions, and the meeting of these equal, opposing currents would stop the flow, producing a result similar to that in the former case.

But if, from difference in the nature of the materials, or in their resistance, or both, molecular motion is more free to extend itself in one of two opposite directions than in the other, and by a transfer of energy, incident to such extension, there occurs a corresponding reduction of such motion in the opposite direction, that is, in electric language, if the current becomes positive in one direction and correspondingly negative in the opposite, it is evident that this motion must extend itself round the circuit continuously, or the current continue to flow from higher to lower potential, so long as the exciting cause continues; the transferred energy, which produces the molecular motion, being again restored to the place of its origin. Just as water in a circular trough, receiving a continued impetus in the same direction at any point, would flow round continuously.

This is precisely what occurs in a battery circuit or in the circuit of an electrostatic machine; materials differing in molecular constitution and resistance, as brass, glass, hard-rubber, pointed conductors and spherical conductors, in the machine, and zinc, fluid, carbon or its equivalent, and copper, in the battery, forming the circuit, which is so arranged in each case that electric action, beginning at a certain point of junction of different materials, is continuously transmitted more easily in one direction than in the opposite; mechanical action being the exciting cause in the machine and chemical action in the battery; and the energy, whether mechanical or chemical, thus absorbed, reappearing as electricity.

We may now consider the application of these principles to thermo-electric generation. In 1821 Seebeck made the discovery that an electric current could be generated by heating or cooling the junction of two dissimilar metals connected in an electric circuit. Seebeck's experiment may be repeated by soldering or fusing together the ends of short pieces of any two metals, differing materially in molecular constitution, as bismuth and antimony, or german-silver and iron, and connecting their free ends, electrically, with a delicate galvanometer. On heating the junction to a temperature above the rest of the circuit, by a spirit-lamp or otherwise, the needle will be deflected, showing the generation of an electric current, and the same effect, with reversed current, will be produced if the junction be correspondingly cooled, as may be done by the application of ice; the direction of the current when the junction is heated being from bismuth to antimony, and, when cooled, from antimony to bismuth; the E. M. F., or potential difference, being in proportion to the difference of temperature between the junction and the other parts of the circuit. Hence, in such a combination, composed of one or more couples, if each alternate junction be heated and the intervening junction simultaneously cooled, the E. M. F. is proportionally increased, the current being from bismuth to antimony across each heated junction, and from antimony to bismuth across each cooled junction, and hence in the same direction round the circuit; and the same would be true of a circuit composed of any other metals having similar molecular differences.

As the capacity of bismuth for heat is much lower than that of antimony, its rise of temperature with the same increment of heat is proportionally greater, and also its fall of temperature with the same abstraction of heat; and as we find that the electric current flows from bismuth to antimony across the heated junction, and oppositely across the cooled junction, it is evident that its flow in each case is from the hotter to the colder metal. But we know that the flow of an electric current is always from higher to lower potential, and in the direction of least resistance, and also that rise of temperature in a metal increases its electric resistance; hence we must infer that, in this case, increase of potential and resistance accompany rise of temperature, and decrease of potential and resistance accompany fall of temperature in each metal respectively, creating a preponderance of both in the hotter metal.

On the molecular theory, it is the propagation of molecular motion, with heat as the exciting cause, which constitutes the electric current; heat undulations being, in some occult manner, transformed into electric undulations. Only a part of the heat supplied is thus

THE RELATIONS OF ELECTRICITY TO HEAT. 263

transformed, the remainder being either radiated, or appearing as heat in elevation of temperature in the circuit; and likewise when heat is abstracted, the remaining heat, set in motion toward the junction by the cooling, is in part transformed into electricity, while the remainder is either radiated, or appears as heat in the reduced temperature of the circuit.

In the above experiment the complete circuit is composed of three metals, copper forming the galvanometer coil and connections, though the generating part, or thermal battery, as it might be termed, is composed of only two metals. But the experiment may be performed with a circuit composed strictly of but two metals. For this purpose let a frame be constructed with a strip of copper or iron, of any convenient length, having its ends bent and soldered to a parallel bar of bismuth; or let it be bent so as to have parallel sides, and its free ends be connected by a cube of bismuth soldered to each. If this frame be mounted on an insulating stand, and a magnetic needle poised on a pivot at its center, the needle will be deflected by heating or cooling one of the junctions, or by heating one junction and cooling the other, as in the former experiment.

No thermo-electric current can be generated in a circuit composed of a single metal of perfectly homogeneous molecular structure; but even with a slight difference, such as may be produced by a coil or twist in a wire, a perceptible current may be obtained, which becomes more marked with increased difference of structure, as between differently manufactured kinds of the same metal, having different degrees of hardness, brittleness, or tenacity: and with continued increase of molecular difference, as between different metals, thermo-electric development increases in like proportion. Which proves that this development is dependent on molecular structure, indicating an intimate relation, if not actual identity, between electricity and molecular motion.

It is found that lead shows no perceptible difference of thermo-electric potential at different temperatures, like other metals; hence it has been chosen as the standard by which the relative thermo-electric potentials of other metals may be compared. In making such comparison the same mean temperature must be chosen for the various metals, since the relative thermo-electric potentials of different metals varies greatly at different temperatures. Taking the microvolt ($\frac{1}{1000000}$ of a volt) as the unit of potential, and $1°$ C. as the heat unit, the following metals, at a mean temperature of $19°$ to $20°$ C., show, according to Matthiesen, the relative thermo-electric potentials indicated, in microvolts, when the temperature of the junction between any two of them is raised $1°$ C. above the rest of the circuit:

Bismuth, Commercial, Pressed Wire............	97
Bismuth, Pure, Pressed Wire...................	89
Bismuth, Crystal, Axial........................	65
Bismuth, Crystal, Equatorial...................	45
Cobalt	22
Mercury.......................................	.418
Lead..	0
Tin..	— .1
Copper, Commercial............................	— .1
Platinum......................................	— .9
Gold	— 1.2
Antimony, Pressed Wire........................	— 2.8
Silver, Pure Hard..............................	— 3
Zinc, Pure Pressed.............................	— 3.7
Copper, Electrolytic............................	— 3.8
Antimony, Commercial, Pressed Wire...........	— 6
Arsenic..	— 13.56
Iron, Soft.....................................	— 17.5
Antimony, Axial...............................	— 22.6
Antimony, Equatorial	— 26.4
Tellurium.....................................	— 502
Selenium......................................	— 807

Thus any two or more of these metals, arranged in this order in a series, would acquire this relative potential difference with a heat increment at the junction, or junctions, of 1° C., and a mean temperature of 19° to 20° C.; each being electropositive to all that follow, and electronegative to all that precede it. The potential difference between bismuth and cobalt, for instance, is $97 - 22 = 75$, and between copper and antimony, $26.4 - 3.8 = 22.6$, while between bismuth and antimony it is $97 + 26.4 = 123.4$; difference between any two above or below zero being obtained by subtraction, while difference between one above and one below is obtained by addition.

In a series composed of any of these metals, arranged as above, the entire potential difference, or thermal E. M. F., is found, as in a battery series, to be equal to the sum of all the differences between each separate couple. Each couple may thus be regarded as a thermal cell, or element, the two metals corresponding to the two electrodes in a battery cell, heat energy, instead of chemical energy, being the exciting cause. Hence in any similar series of metals, $ABCD$, the sum of the potential differences between each couple, as A and B, B and C, C and D, is the same as the potential difference, or E. M. F., between the extremes A and D; so that if a direct junction were made between A and D, and the intervening couples omitted, the E. M. F. would be the same as in the full series, as may be verified numerically in any series chosen from the table.

In a circuit composed of several couples of any two metals, alternate junctions require to be either heated or cooled, or each alternate junction heated and the intervening one simultaneously cooled, as the heating or cooling of two adjacent junctions to the same temperature would produce opposite, neutralizing currents.

The table is not intended to embrace all the substances which manifest electro-thermal properties, but only a few of the metals in which those properties are prominent; such properties being common to a large class of bodies, both metallic and non-metallic. The potential differences given must be understood to apply only in a general sense, as differences of molecular structure produce, as shown, great variation in this respect; so that in different experiments the results obtained from the same metals procured from different sources would be only approximately the same.

Thermo-Electric Diagrams.—Sir William Thomson proposed a graphic representation of the relative thermo-electric potentials of different substances at different degrees of temperature, consisting of a diagram having vertical lines representing the differences of temperature, and lines approximately horizontal representing the thermo-electric differences of potential, and in 1856 used such a diagram for the first time to illustrate a lecture. Fig. 90 shows a diagram by Prof. Tait constructed in this way. Lead, being thermo-electrically constant at different temperatures, is represented by a perfectly horizontal line, marked o, while the other metals are represented by lines tilted at the various angles required to show their relative thermo-electric differences, at different temperatures, with respect to lead, and hence with respect to each other. Lines representing metals whose potential difference with respect to lead increases negatively or decreases positively with increase of temperature descend from left to right; while those representing metals whose potential difference increases positively or decreases negatively with increase of temperature ascend from left to right. Thus zinc, at 19° to 20° C., is shown to be about −3.7 below

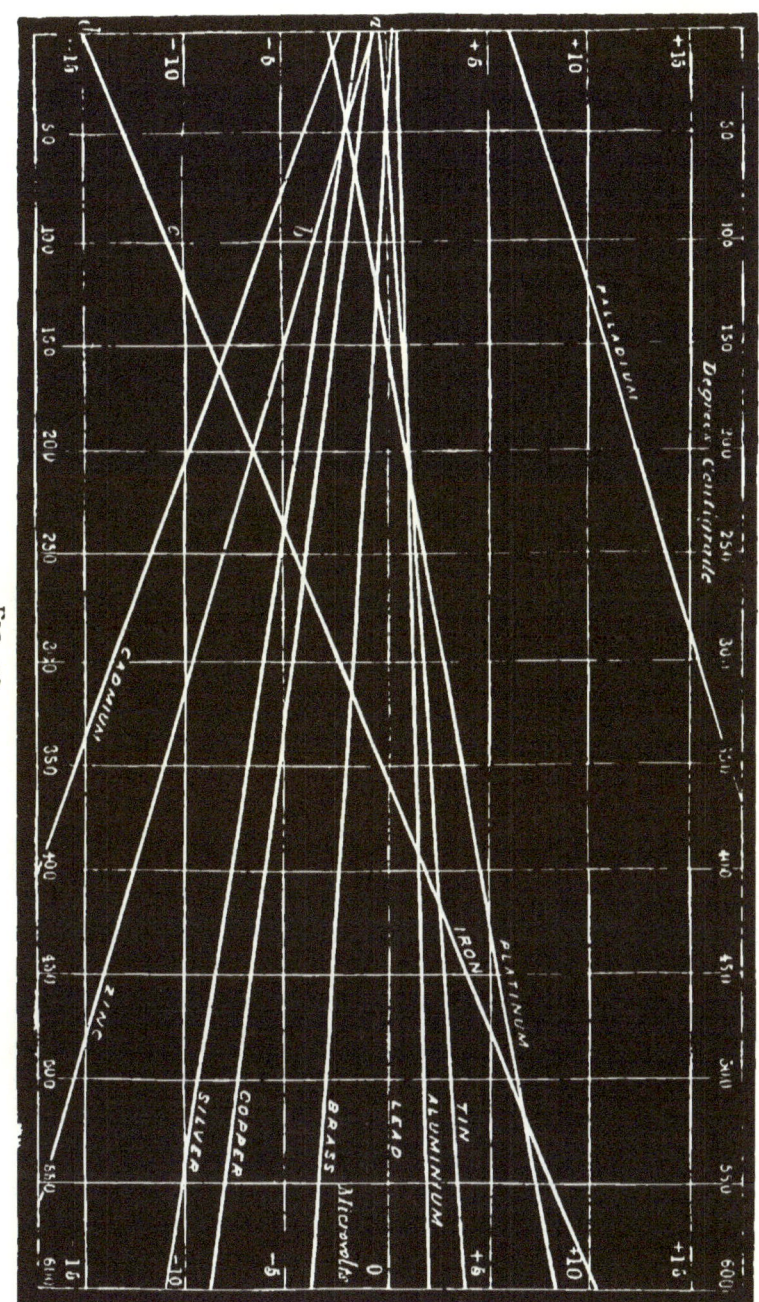

FIG. 90.

lead, as given in the table, while at 480° C. it is —15' below.

Thermo-electric differences may be represented also by the areas formed by the lines. Thus in a zinc-iron couple, with one junction at 100° C. and the other at 0° C., the thermo-electric difference is represented in the diagram by the area a, b, c, d. For small temperature differences, of one or two degrees, the superficial contents of the areas are practically the same as between rectangles, and hence are obtained by simply multiplying the temperature differences by the potential differences; but for large temperature differences the irregular shape of the areas requires special calculation in each case.

The Peltier Effect.—Peltier's observations on the thermo-electric circuit led him to the natural conclusion that if difference of temperature at the junctions could generate an electric current, then, conversely, the passage of an electric current through such a circuit must generate a corresponding difference of temperature at the junctions, and experiments made by him in 1834 verified this conclusion. Hence this phenomenon, which is now a well-established thermo-electric law, has been called the *Peltier effect*, in distinction from the generation of heat by the resistance of the circuit, as observed by Joule, which is known as the *Joule effect*.

These two effects are entirely consistent with each other and may occur simultaneously in the same circuit. For, in any circuit, whether composed of one metal or several, the temperature varies in proportion to the *square* of the current's strength, in accordance with the Joule effect; but if the circuit is composed of different metals, or different kinds of the same metal, there occurs also a transfer of heat from one junction to another in proportion simply to the current's strength, so that one junction is heated while the other is correspondingly cooled, in accordance with the Peltier effect. In the Joule effect the amount of heat generated in the circuit, as a whole, is not varied by the direction of the current, while in the Peltier effect the transfer of heat is reversed by reversal of current; so that junctions heated by a current flowing in a given direction, as from antimony to bismuth, are correspondingly cooled by the same current flowing in the opposite direction, as from bismuth to antimony, while the alternate junctions cooled in the first instance are correspondingly heated in the second. The reduction of temperature in a bismuth-antimony combination may thus become so great as to freeze water in a cavity at the cooled junction.

Thermo-Electric Inversion.—Prof. J. Cumming found that, in a copper-iron couple, iron ceases to be electronegative to copper when the temperature of the junction is raised to 280° C.; the current from copper to iron ceasing, and the Peltier effect also disappearing, when a current is transmitted in either direction. But when the temperature of the junction is raised above 280° C., iron becomes electropositive to copper and the Peltier effect is also renewed. This is illustrated by the iron and copper lines in the diagram which cross each other at the neutral point, iron being represented as electronegative to copper on the left of this point and electropositive on the right: similar thermo-electric inversion being also shown in other metals.

The potential differences given in Tait's diagram vary somewhat from those given in Matthiesen's table, with which Cumming's experiments seem to accord more closely; only approximate accuracy being attainable in different experiments, for the reason already given.

The Thomson Effect.—This thermo-electric inversion

led Sir William Thomson to conclude that since there is no heat development, or Peltier effect, at the junction of a copper-iron couple, at 280° C., by the passage of an electric current through it, therefore, conversely, there can be no accumulation of heat at this point, at like temperature, when the current is generated by heat supplied to it, and therefore the heat supplied must be absorbed by other parts of the circuit than the junctions, and hence must pass between differently heated parts of the same metal.

Experiments with different metals verified this conclusion, showing that when a thermo-electric current passes through a conductor, from a hotter to a colder part, there is a transfer of heat, which in some metals, as copper, is from the hotter to the colder part, while in others, as iron, it is from the colder to the hotter part: but when the direction of the current is from a colder to a hotter part this transfer is reversed. This electric convection of heat in the same metal is known as the *Thomson effect*, in distinction from the Peltier and Joule effects.

It follows from the above that in a copper-iron circuit, when a current is generated by heating a junction to any temperature below 280° C., the current being from copper to iron is from cold to hot in the copper and from hot to cold in the iron, so that in both metals heat is transferred to the junction; but when the temperature of the junction is raised above 280° C., the current being reversed, the heat transfer is also reversed, and is from the junction in both metals: while with the junction at 280° C., there being no current, there is no transfer of heat in either direction. And the same principles apply to any circuit having similar thermo-electric inversion.

The Thermopile.—It has not yet been found possible

to construct electric generators of general practical efficiency on the principle of the direct conversion of heat into electricity. Generators of this kind, constructed by Clamond and others, have not fulfilled the hopes raised by their first apparent success; the generation of strong currents, combined with the heat necessary to produce them, seems to effect, in a short time, such permanent change of molecular structure as to reduce the production and maintenance of potential difference between the different metals below the point of practical efficiency.

The difficulties in such construction become further manifest when we consider that comparatively few of the metals given in the table are practically available for this purpose, either in consequence of small potential difference, extreme rarity, as in the case of tellurium and selenium, volatility when heated, or other cause. Of the available metals, bismuth and antimony have the highest potential difference, and can be used at moderate temperatures; bismuth melting at $264°$ C. and antimony at $450°$ C.

The *thermopile*, represented by Fig. 91, is constructed with a number of small, short metal bars, usually of bismuth and antimony, arranged side by side in couples, junctions being formed between each pair of ends in alternate order, by soldering or fusing; the arrangement being such that the current must pass from one metal to the other through the entire series. By having as many layers as there are bars in a layer, a compact, cubical form is obtained.

The bars thus arranged and properly insulated are inclosed in a brass case, open at the ends, and mounted on a stand provided with apparatus for elevating and adjusting them to any required height or angle. Conical caps are fitted to the ends to admit the heat radiated from any object whose temperature is to be tested, and to exclude radiation from other sources; and binding-screws are provided for the galvanometer connections.

The alternate junctions being at opposite ends, one set may be exposed to heat while the alternate set are cooled; and the entire potential difference, or E. M. F., being equal to the sum of the potential differences in

FIG. 91.

the series, a very sensitive apparatus, for investigating slight differences of temperature, may be obtained, when the instrument is used in connection with a sensitive galvanometer; the needle responding instantly with a prominent movement, easily read from the scale, to temperature differences hardly perceptible in the thermometer: the heat generated by the bending of a copper wire being sufficient to produce a deflection of several degrees. The highly important investigations of Melloni and of Tyndal on heat were conducted with the aid of such an apparatus.

The relative E. M. F. of the thermopile as compared with other generators is very small. Taking 200 microvolts as the average E. M. F. attainable by the simultaneous heating and cooling of the opposite junctions in a single bismuth-antimony couple, the total E. M. F. in a thermopile, or multiplier, of 50 such couples would be $50 \times 200 = 10{,}000$ microvolts, or $\frac{10000}{1000000} = \frac{1}{100}$ of a volt; so that the combined E. M. F. of 100 such generators would be only equal to that of a single Daniell cell. But as the comparative resistance of the thermopile is also small, the current is comparatively large: supposing the resistance in the above case to be $\frac{1}{10}$ of an ohm, then $\frac{\frac{1}{100}E}{\frac{1}{10}R} = \frac{1}{10}C$, or $\frac{1}{10}$ of an ampere. In the largest Clamond thermo-electric batteries, consisting of 150 iron-galena elements, the estimated E. M. F. is only $5\frac{4}{10}$ volts, and the internal resistance 2 ohms, which would give a current of $2\frac{7}{10}$ amperes, in an external circuit of no resistance. A Daniell battery of the same number of elements similarly joined, in series, would have an E. M. F. of about 150 volts and an internal resistance of about 300 ohms, which would give a current of $\frac{1}{2}$ an ampere, in a similar external circuit, less than $\frac{1}{4}$ that of the thermo-electric battery, though its E. M. F. is 28 times as great.

Electric Welding.—This highly important application of electricity has been largely developed by Prof. Elihu Thomson since 1886, and has now attained a wide range of practical work. It consists in uniting pieces of metal by pressing them together, end to end, and heating the juncture by an electric current till the metal becomes sufficiently plastic to form a perfect joint; only so much of it being included in the circuit as may be necessary for this purpose.

The alternating current is employed, and applied by

the welder shown in Fig. 92, which consists of a converter and clamping apparatus combined. The converter, shown in the rear, is constructed with a laminated iron core inclosing a massive copper tube, equivalent

FIG. 92.

to a single coil, which forms the secondary circuit. The primary circuit consists of an insulated copper coil wound in two sections through the interior of this tube, as shown, and inclosing its upper and lower parts together with the adjacent parts of the core. This circuit is connected with the dynamo by the terminal

THE RELATIONS OF ELECTRICITY TO HEAT. 275

wires shown in the rear, and the secondary circuit is connected, on the right, with a massive grooved copper bar, to which is fitted the copper sliding-block A. Two massive copper clamps, C and C, grasp the two bars to be welded together, the right one movable in connection with the sliding-block A, and the left fixed; and to this fixed clamp the secondary circuit is connected on the left.

Pressure being applied to the block A by the crank B and connected gearing, the right bar to be welded is forced against the left; the circuit being opened and closed by a switch connected with a treadle, and the current regulated by a reactive coil connected with the primary circuit.

A dynamo, specially adapted to this work, furnishes a current which, in the 20,000-watt welder, has an E. M. F. of 300 volts at the terminals of the primary circuit, which is reduced, in the secondary circuit, to about 1 volt; and the efficiency being about 80 per cent, the maximum current is about 16,000 amperes. The welding capacity of a welder of this size, for bar-iron, ranges from bars $\frac{3}{8}$ of an inch in diameter to bars of $1\frac{1}{4}$ inch diameter; the range for brass being three fourths of this, and for copper one half.

To adapt the welder to different kinds of work, its primary circuit is connected in series to an auxiliary converter of special construction, by which the E. M. F. can be more fully controlled. The primary circuit of this converter is wound on a section of a laminated iron core, composed of a split ring, the slit being on the opposite side from the coils, which are so arranged that they can be joined either in series or parallel by a switch. The core incloses an iron armature, upon which is mounted the secondary circuit, consisting of a massive brass casting, which also includes a section of the core and may be rotated so as to include either the primary circuit or the slit, as required. When rotated so as to include the primary, the E. M. F. is reduced to one half that which it is when the secondary is opposite the slit, across which no lines of force can pass, and where the magnetism is therefore at its minimum. The E. M. F. can also be either increased or diminished by joining the coils of the primary, either in series or in parallel; hence its variation in the welder, by these various means, includes a wide range.

The above, known as the indirect method, is employed for the heavier and more complicated kinds of welding, and where several welders are operated by current from a single dynamo; while for the lighter, simpler kinds the direct method is employed, in which the welder is connected directly to the dynamo, the armature of which has a high potential circuit of fine wire, in series with the field-magnet coils, which acts inductively on a low potential circuit, composed of a massive, U-shaped, copper bar, connected directly to the welding apparatus, the construction of which is the same as already described.

The direct current may be employed, but the alternating is preferred on account of its higher efficiency and freedom from electrolyzing effects, a point of special importance in the welding of alloys.

The ends to be welded together are rounded so that contact shall be first made at the center, and the weld being from the center outwards, oxidized particles and other impurities are forced out as the ends are pressed together, making a perfect joint, superior to any which can be made by forging; and the entire process being thus open to the inspection of the operator, flaws cannot escape observation. Manual pressure can be employed in ordinary cases, but for more difficult welding, where great accuracy is required, pressure by hydraulic or other mechanical power is preferred.

The greatest heat is developed at the center of the weld, extending only a short distance on either side, and varying directly as the resistance, which increases with the rise of temperature. Bars of inch iron become red hot for a distance of $1\frac{1}{2}$ inches on either side of the weld, but are comparatively cool at a distance of $2\frac{1}{2}$ inches; and the operation being completed in 40 seconds, the time is too short for diffusion of the heat by conduction; hence waste of energy from this source is reduced to the minimum. The time varies for metals of different kinds and sizes, from 1 or 2 seconds for fine wires to 2 or 3 minutes for heavy bars; wrought-iron bars, 2 inches in diameter, requiring an average time of about 97 seconds; $2\frac{1}{2}$-inch iron pipes, $\frac{1}{4}$ inch thick, 61 seconds; the average time for copper bars being about $\frac{5}{8}$ that required for wrought-iron bars.

The E. M. F. is so low that the enormous current required for heavy work is perfectly safe; and conductors carrying currents of many thousand amperes, but having an E. M. F. of only a fraction of a volt, may be handled with impunity and without sensible effect.

The range of application is almost unlimited, embracing not only all welding hitherto considered practicable, but a large amount considered either wholly impracticable or extremely difficult, ranging from the most refractory metals to alloys fusible at 90° C. Not only can such metals as cast-iron, copper, lead, tin, zinc, brass, german-silver, and bronze be welded, each to its own kind, but any of these dissimilar kinds can be welded together. Steel cables composed of a large number of fine wires, tubing, and various kinds of metal work usually united by screws, rivets, soldering, or brazing, can be welded by this process;

also articles which, from their peculiar shape, are difficult or impossible to weld in the ordinary way. It has also a highly important application in the expeditious repairing of broken machinery on ships, in factories, and elsewhere.

Welds made by this process have been subjected to the severest practical tests by the United States naval authorities and various, eminent, civil and electrical engineers, and have received their unqualified approval for superior strength and tenacity.

CHAPTER XI.

THE RELATIONS OF ELECTRICITY TO LIGHT.

The Relations of Electric Heat to Electric Light.—It has been shown that heat is always a result of electric resistance, and is in proportion to such resistance; and as a certain degree of resistance is found in every conductor, it follows that heat always accompanies electric transmission. When the heat increases to a sufficient degree of intensity, light is produced, either by incandescence or combustion according to the nature of the medium of transmission. Hence the electric generation of light follows that of heat and is dependent on heat intensity; heat being produced without light, but light never being produced without heat.

Heat and light, according to well-established theories, being considered different modes of molecular motion, if electricity also be so considered, the difference between the three would seem to consist in the nature of the motion in each case, and may be attributed to differences in the length, amplitude, rapidity or phase of undulation peculiar to each, as pertaining both to the molecules of the conductor, and to the medium of transmission through space.

Neither phenomenon is developed at the expense of the other, except as the nature of certain conductors produces variation between the development of heat and electric current; hence if these are different kinds of molecular motion, they must occur in such a manner as not to neutralize each other. It has been shown how such different kinds of motion may coexist without interference in the magnet, and similarly here, motion whose general direction is in lines, straight, curved, or spiral, would not interfere with transverse undulations, nor would either interfere with rotary motion of the molecules.

It is not impossible that two or more of these phenomena may be identical; that the heat undulations, or the electric undulations, or both, are, at a certain degree of intensity, recognized as light; though, in the present state of our knowledge, it is more in accordance with observed facts to assign to each a distinct mode of motion; that of heat being comparatively slow, with considerable length and amplitude of undulation, while those of light and electricity are inconceivably rapid, with undulations of a corresponding character.

We have seen that heat and electricity reproduce each other directly, but that in the electric production of light, heat intervenes, and that the light is apparently a result of the heat rather than of the electricity; for when the heat is produced by any other method, light usually follows increase of heat intensity in the same manner, though we have no direct evidence of the presence of electric action: and yet it is not impossible that electricity, though occult, may be present as an active agent, or that the light and the electricity may be identical.

Photo-Electric Generation.—While the direct generation of light by electricity is not clearly apparent, the direct generation of electricity by light has been effected experimentally, though it has not yet been found possible to construct practical generators on this principle.

The first experiment of this kind was made by Becquerel about 1850, who found that when one of two silver plates, freshly coated with silver chloride and immersed in water, is exposed to light, an electric current, indicated by a connected galvanometer, flowed to the exposed plate from the opposite pole.

In 1875-6 Adams and Day, English electricians, made a very extensive series of experiments to ascertain the electric relations of selenium to light; one result of which was the discovery of electric generation by this metal under the influence of light. A small piece of selenium, whose electric resistance had been reduced by annealing, had platinum terminals fused into its opposite ends; the platinum wire being formed into little rings on the inserted ends, to giver fuller contact. On exposure of the selenium to candle-light the passage of an electric current was indicated by a prominent deflection in a connected galvanometer; the direction of the current being from the part least exposed to the part most exposed, a result similar to that in Becquerel's experiment.

That this was not a thermo-electric current was proved in various ways: 1. The current began promptly with the exposure, and ceased promptly with the exclusion of the light, instead of showing the more gradual increase and decrease of current due to heating and cooling. 2. The current, in most of the experiments, was the result of exposure of the body of the metal, while the thermo-electric current results from exposure of the junctions. 3. When the light was focused on a junction the direction of the current was from selenium to platinum, while that of a thermo-electric current would have been from platinum to selenium; this direction being also, as will be perceived, from the least to the most exposed part of the selenium, as before.

In 1887, Prof. Edlund constructed a generator by melting a very thin layer of selenium on a disk made of a metal with which it could unite chemically, and covering this layer with gold-leaf made so thin that the sunlight could penetrate to the selenium. Connection with a galvanometer being made between the gold-leaf and lower disk, an electric current was developed on exposure to the sun's rays, which responded promptly to the influence of the light, and ceased promptly with its exclusion, thus proving its photo-electric character, as in the former example.

Photo-Electric Reduction of Resistance in Selenium.—The electric resistance of ordinary, vitreous selenium is $3.8 \times 10^{10} = 38,000,000,000$ times that of copper, but when annealed by being kept for several hours just below the point of fusion, $220°$ C., and then cooled slowly, it becomes crystalline and its resistance is materially reduced. The difference of crystalline structure produced by the more rapid cooling of the exterior than the interior has been assigned by Gordon as a probable reason for its property of photo-electric generation.

It was found by Adams and Day that the resistance of this annealed selenium, when a battery current is passed through it, is much less in the light than in the dark; *the resistance varying directly as the square root of the quantity representing the illumination.*

Bell and Tainter utilized this property of selenium in the construction of their *photophone*. A narrow strip of selenium connected at the edges with broad plates of brass furnishes a photo-receiver of large surface exposure and of low resistance as respects form; the selenium furnishing a resistance highly sensitive to light and varying under its influence from 300 ohms to 150. This photo-receiver being placed in a battery circuit connected with a telephone receiver, the varying light reflected from a distant point by a thin mirror, constituting the disk of a telephone transmitter, which responds to the undulations of the voice, produces corresponding variations in the battery current, by which the voice is reproduced in the telephone receiver, as explained in connection with the telephone.

Tellurium has the same photo-electric properties as selenium in less degree, and carbon also shows similar properties.

Polorization of Light.—The ether undulations, supposed to constitute light, are believed to be transverse to the direction of the rays. This transverse undulation is supposed to be equal in all directions within a circular space, so that the theoretical conception of a ray viewed endways in cross-section would be that of a circle composed of an infinite number of planes of undulation in which the undulations, by mutual adaptation, occur without interference. As if numerous fine wires, each bent into short curves, in the same plane, at right angles to the wire's length, were fitted together so as to form a long slender cylinder, with these curves crossing each other at all possible angles along its central axis.

But under certain conditions of transmission and reflection, the ray becomes flattened, as if compressed between opposite lateral forces, so that these undulations all occur in one plane, and the ray is then said to be *polarized*.

This happens when light is transmitted through certain crystals, especially tourmaline. If two thin plates of tourmaline be placed with their surfaces parallel to each other, and a ray of light be transmitted through them at right angles to their surfaces, and to a certain direction in each, known as its optical axis, the light will pass freely through both to a screen beyond, so long as these axes are parallel. But if either crystal be turned so that the optical axes are at an angle, the surfaces being still parallel, the light which passes through one is obstructed in the other, gradually disappearing from the screen as the angle increases, till at 90° it is entirely extinguished. If the rotation be continued in the same direction, the light gradually reappears on the screen, and regains its original brightness when the axes again become parallel. The crystal on which the light is first received is known as the *polarizer* and the other as the *analyzer*.

The theory of this phenomenon is that the undulations in passing through the polarizer are changed from the phase of a circle to that of a plane, in which form they readily pass through the analyzer so long as the optical axes of both crystals lie in the same plane; but when the planes of the axes cross, it is as impossible for the polarized light to pass through the analyzer as it would be for a metal rod, compressed into a sheet between rollers, to pass crossways through the wires of a bird-cage.

Light when reflected at certain angles from certain substances becomes polarized as well as when transmitted, the polarizing angle varying according to the nature of the reflecting substance. The analyzer in this case may be either a reflector or a transmitter, and the polarized ray is reflected, transmitted, or extinguished according to the angle at which it meets the analyzer.

Magneto-optic Polarization.—Faraday's Discoveries.—In a series of experiments, made in 1845, Faraday found that polarized light is influenced by the electro-magnetic current. A polished piece of "heavy glass"—silicated borate of lead—about 2 inches square and $\frac{1}{2}$ an inch thick, was interposed edgeways in the path of a ray of lamp-light, polarized by reflection from a plane glass surface; the analyzer being turned so as to extinguish the ray. A U electromagnet was placed close to the glass, in such position that a line through its poles, which were about 2 inches apart, was parallel to the direction of the ray. On the passage through its coils of an electric current from a battery of five Grove cells, the extinguished ray again passed through the analyzer, proving that its plane of polarization had been rotated into a new position by the electromagnetic action; which was confirmed by the fact that, by a further rotation of the analyzer, an angle was found in which the magnetized ray was extinguished, but in which the ray was transmitted when no current was passing—a reversal of the conditions of transmission and extinction found in the first position.

Faraday found that, to produce these results, a solid or a liquid medium of transmission was necessary for the reception of the magnetic action, but failed to obtain them by such action on air or other gaseous medium, or in vacuo. He also found that the direction in which the plane of polarization was thus rotated coincides with that in which the magnetizing current passes round the magnet; reversal of current consequently producing reversal of this rotation. But it was subsequently ascertained by Verdet that this coincidence of direction is true only of diagmagnetic bodies, while, in certain paramagnetic bodies, this rotation is opposite to the direction of the magnetizing current.

It should be especially noticed that the direction of the magnetic lines of force, from pole to pole, was, by the position given to the magnet, made parallel to the ray.

Faraday varied his experiments by using different kinds and different forms of magnets, and placing the glass, or rather medium, in different relative positions; but to obtain the effect described, the parallel position of the ray to a line through the poles was requisite. He also used a pair of bar electromagnets with tubular cores, so placed that a ray could be transmitted through both and received on any medium placed between dissimilar poles, which, as in the U magnet, were about 2 inches apart.

Passing the ray horizontally across a *single* pole, with the magnet in a horizontal position, he found the ray's rotation, when the glass was on the side next the analyzer, to be the reverse of what it was when the glass was on the opposite side; change of pole or reversal of current producing reversal of rotation. But, with the glass above, below, or in front of the pole, no rotary effect was produced. The cause of these various effects becomes obvious when we consider that the lines of magnetic force radiate in all directions from a single pole: hence, when the glass was in the horizontal plane of the magnet, these curved lines, in that plane, were nearly parallel to the short portion of the ray transmitted through the glass, but radiated in opposite directions on opposite sides of the pole; so that on one side they coincided with the direction of the ray's transmission, and on the opposite side were opposed to it; but above or below the pole they were at right angles to the ray, while in front of it radiation was equal in opposite directions.

Another rule given by Faraday for finding the direction of the ray's rotation, with diamagnetic bodies, which has a special application to the case of a single pole, is substantially as follows: A ray of light, coming to the observer, is rotated in the same direction as watch-hands move. when the magnetic lines of force parallel to it are radiated from a north pole in the *same* direction as the ray, or from a south pole in the *opposite* direction; reversal of the ray's direction producing reversal of rotation.

Faraday obtained the same effect, in a limited degree, from steel magnets as from electromagnets; also from coils without iron cores; proving that the effect is chiefly magnetic, though also electric. He also found that this effect is independent of any specific polarizing property normally pertaining to the diamagnetic body through which the ray is passed; the electromagnetic polarizing effect being either increased or diminished by such specific property, according as it produced rotation in the same or in the opposite direction. He could not produce any change in this effect by any degree of motion given to the dielectric while under the joint influence of magnetism and light. He noticed that the rotation increased slowly, requiring about two seconds after the closing of the circuit for the attainment of the full effect, but that it ceased promptly on opening the circuit. The first result he attributes to a lag in the magnetic saturation of the core, while the second showed the intimate relation of this effect to electromagnetic action. His conclusion in regard to magnetic lag was confirmed by the fact that there was no lag when the coil alone, without a core, was used; the rotation responding promptly both to the opening and closing of the circuit. He also found that any addition made to the dielectric on either side, and not in the line of the ray, produced no difference in the rotary effect.

His final conclusion is, that since this effect is essentially the same in character under all these varying conditions, and is independent, in this respect, of the nature of the dielectric, or its own specific rotative force, therefore the magnetic force and the light have a direct, mutual relation, but require the intervention of matter as the medium of action.

Verdet's Discoveries.—Experiments made by Verdet in 1852 confirmed the results obtained by Faraday, except in regard to the direction of the rotation produced by certain paramagnetic bodies, as already explained. His apparatus consisted of two powerful electromagnets with hollow cores, similar to those used also by Faraday, through which light could be transmitted to the medium interposed between dissimilar poles, parallel to the lines of force. He also used a U electromagnet with massive, slotted pole-pieces, through which the light could be transmitted at any desired angle; the magnet having also a rotary movement by which, the angle between the lines of force, from pole to pole, and the ray could be adjusted and measured with a graduated scale and vernier. The principal substances used as media were the "heavy glass," used by Faraday, common flint glass, and carbon bisulphide.

Verdet endeavored to ascertain not only the facts in regard to electromagnetic polarization, but also the laws which govern it; and to determine the specific electromagnetic rotative force of different substances. His principal deductions are embodied in the following law: *The rotation of the specific electromagnetic plane of polarization for any substance is directly proportional to the strength of the magnetic action, to the thickness of the medium traversed jointly by the magnetism and light, and to the cosine of the angle between the ray and the lines of magnetic force.*

Verdet chose water as his standard of comparison for specific rotative differences; but Gordon, who subsequently made a special investigation of this subject, found carbon bisulphide a more reliable standard. Hence taking the specific magneto-rotative force of this substance as unity, that of water is found to be 0.308 and that of "heavy glass" 1.422.

Becquerel's Discoveries.—These are the specific differences for white light; but this force has been found to vary for different colored rays, and since difference of color is believed to be due to difference of wave-length, H. Becquerel, who, in 1880, made a special investigation of this branch of the subject, claims to have found that *the rotations of different colored rays vary* (very nearly) *in the inverse ratio of the squares of their wave-lengths.* Thus taking the rotation produced by carbon bisulphide in green light as unity, that produced in red light is 0.6 and in blue light 1.65. The wave-lengths assigned to each, in ten-millionths of an inch, being 211 for green light, 256 for red, and 196 for blue, if each number be divided by 211 and the quotients squared, the reciprocals of the squares, expressed decimally, correspond approximately to the respective rotations given above, in accordance with Becquerel's law.

Kündt and Röntgen's Discoveries.—In 1879, Kündt and Röntgen, with a 65-cell Bunsen battery, and electromagnets wound with 2400 turns of wire, discovered the magnetic rotative force of air and other gaseous bodies, which Faraday with a 5-cell Grove battery failed to discover. They found that air, oxygen, nitrogen, carbonic acid, coal-gas, ethyl, and marsh-gas, all rotate the ray in the direction of the magnetizing current, like water and carbon bisulphide; that the degree of rotation, which is very slight, varies greatly in differen* gases, and is proportional in each to the density of the gas; and that light, traversing the atmosphere in the plane of the magnetic meridian, is rotated, by the earth's magnetism, 1° for every 316 miles of air traversed.

Becquerel, whose experiments were made a year later, found that the rotation of oxygen is opposite in direction to that of the other gases mentioned; such difference in observation being easily accounted for by the small degree of the observed rotation.

Kündt discovered, in 1884, that light transmitted through a film of iron, of such tenuity as to be transparent, is rotated in the direction of the magnetizing current, as in diamagnetic bodies.

Kerr's Discoveries.—In 1875 Dr. Kerr discovered that

light, polarized in a plane, when transmitted through a dielectric, at certain angles, under intense electric strain, suffers double refraction and is changed into that mode of polarization known as elliptical, in which the undulations occur in two planes crossing each other at right angles.

For this purpose he used a rectangular prism of plate glass, in which holes were drilled at each end to within $\frac{1}{8}$ of an inch of each other at the center, into which were inserted the wire terminals of a powerful induction coil. A receptacle of similar shape, and of special construction, was also provided for experiments on various liquid dielectrics, as carbon bisulphide, benzol, paraffine oil, kerosene, oil of turpentine, and olive oil.

The light, after passing through a polarizing crystal, was transmitted through the dielectric at right angles to the direction of the wires; the polarizer being turned as required to cause the plane of the polarized ray to form with this direction any angle desired; and the ray, thus transmitted, was received by the analyzer.

This will be better understood if the dielectric be conceived as lying across this page, the direction of the wires being the same as that of the printed lines, and the ray, polarized in a plane, transmitted at right angles to the surface of the paper; the plane of the ray being turned so as to form an angle with the printed lines; as if a thin knife-blade, turned at an angle to the lines, were thrust through the paper.

The ray being thus transmitted, and the analyzer turned so as to extinguish it, reappeared, on the passage of the current, when the electric strain reached a high degree of intensity; being brightest when the plane of the ray was at an angle of 45° to the direction of the wires —or electric strain—but becoming dimmer as the angle either increased or diminished; and being extinguished when the plane of the ray was either parallel to the direction of the electric strain or at right angles to it.

Dr. Kerr's conclusion from these experiments is, that, in any given dielectric, *the quantity of this optical effect— or intensity of electro-optic action—per unit of thickness of the dielectric, varies directly as the square of the resultant electric force produced in the dielectric.*

In 1877 Dr. Kerr discovered that light reflected from the end of an electromagnetic pole having a polished surface, is rotated in a direction opposite to that of the magnetizing current, and hence in opposite directions by dissimilar poles.

In order to concentrate the magnetic force on the polarized ray, he used a block of soft-iron which he called a "submagnet," having a rounded angle which was placed within $\frac{1}{20}$ of an inch of one pole of a U electromagnet. The ray, polarized in a plane either parallel or perpendicular to the plane of the angle of incidence, met the pole's surface in this narrow space, and was thence reflected to the analyzer, through which it passed when magnetized, being rotated as above, but by which it was extinguished when not magnetized. When the ray was polarized in a plane forming an oblique angle with the plane of the angle of incidence, the magnetism produced elliptic polarization, as in transmission through a dielectric under electric strain, and the ray could not be extinguished as before. The angle of incidence is that included between the incident ray and a perpendicular to the reflecting surface; its plane being known as the plane of incidence.

Dr. Kerr also found that when polarized light is reflected from the side of an electromagnet, the resulting rotation, except under certain conditions, is in the same direction as that of the magnetizing current.

In this investigation he dispensed with the submagnet,

and he used, for a reflector, the side of a soft-iron armature, laid across the ends of the poles of a U electromagnet. The ray, received through a slit in a screen, passed through the polarizer, and was reflected to the analyzer from a side of the armature perpendicular to that across the poles, in a plane at right angles to the magnet's plane.

When the ray was polarized in a plane parallel to that of the angle of incidence, the rotation was in the same direction as that of the magnetizing current, for any angle of incidence; but when polarized in a plane perpendicular to that of the angle of incidence, the rotation was in this direction only for angles of incidence between $75°$ and $80°$, and in the opposite direction for angles between $75°$ and $30°$.

Effects of Double Reflection.—It has been observed that when light is polarized by reflection from a plane surface, a second reflection, in the opposite direction, from a parallel plane surface, at the same angle and in the same plane, annuls ordinary polarization but doubles magnetic polarization. Hence, with ordinary polarization, an even number of such reflections annuls, while an odd number gives the same amount as a single reflection: but, with magnetic polarization, the effect, under these conditions, is multiplied by the number of reflections.

Summary.—The results of all these various observations, in which are comprehended about all that is known of the relations of electricity to light, may be briefly summarized as follows:

1. Light can be generated by electricity and electricity can be generated by light.
2. Polarized light, transmitted through a dielectric, has its plane of polarization rotated either by electromagnetic force, by magnetic force alone, or by the force

of an electric current alone, in the same direction as the current which produces, or would produce, the resulting magnetism.

3. Polarized light, reflected from the *end* of an electromagnetic pole, has its plane of polarization rotated in a direction opposite to that of the magnetizing current, when polarized either parallel or perpendicular to the plane of incidence. But when reflected from the *side* of an electromagnetic armature, the rotation is, for nearly all positions of polarization, in the same direction as that of the magnetizing current.

4. Light transmitted through a dielectric under electric strain undergoes double refraction when polarized at an angle of 45° to the direction of the strain.

5. Reflection which annuls ordinary polarization multiplies magnetic polarization.

Maxwell's Theory.—It has been already suggested that magnetism may be a mode of molecular or other motion having the phase of a vertical whorl around a central axis of propagation. This is the theory of Clerk Maxwell, in which he attributes magnetism to an undulatory motion of this kind in the ether. Applying this theory to the magnetic polarization of light, he conceives that the polarized ray, passing through the magnetic field, has its plane of polarization rotated into a new angle, in this magnetic whorl, in which it can pass through the analyzer, where it was before extinguished.

This theory certainly accounts in a very satisfactory manner for the opposite phases of rotation produced by opposite poles, and otherwise, under the various conditions of transmission and reflection which we have been considering. For if such a vertical whorl exists in the magnetic field, it is evident that the rotation of the polarized ray, in passing through it, would depend on the angle between the plane of the ray and that of the whorl; so that the different phases observed to exist are just those which should result from such conditions.

Molecular Theory.—It is not improbable that these phenomena may be due to modes of molecular motion, magnetic or electric, in the substance of the media, rather than to undulations of the hypothetical ether; such a theory being as consistent with the various effects observed as that of the undulating ether. The rotation produced by reflection of the ray from a magnet is no exception to this; the molecular motion of the reflecting surface producing the rotation, which is intensified by the passage of the ray through the magnetic field having the air for its medium, to which the molecular motion of the magnet is communicated.

Strain in the Media.—It is evident that the rotation of the ray, and the other effects observed, seem to result from magnetic or electric strain in the media rather than in the light itself, and that the effect on the light is secondary: still it is none the less evident that these effects are as truly modes of polarization as the polarization which occurs in the ordinary way in the crystal; the latter being, as we have seen, due to the peculiar crystalline structure, by which the undulations are all forced into the same plane, while, in the former, the structure of the media, solid, liquid, or gaseous, changed by magnetic or electric action, forces this plane into a new angle.

The motion, given the media by Faraday, would not disprove this, since it is probable that the magnetic action would produce change of structure in the medium in each new position much more rapidly than the mechanical action could produce change of position; so that the direction of the strain would be the same as if the medium were stationary.

Quincke attributes the double refraction obtained by Dr. Kerr to an electrostatic strain producing either expansion or contraction in the media according to the substance employed. Fontana noticed that the Leyden jar becomes slightly expanded when charged; an effect attributed by Volta, Priestley, and Duter to electric compression of the glass.

Electric Lighting.—It has been shown that when the heat developed in a conductor by its resistance attains a sufficient degree of intensity light is produced; and on this principle, by the use of conductors of high resistance, we obtain the electric light, either as the result of incandescence or combustion.

The Arc Light.—The electric light was discovered in 1813 by Sir Humphry Davy, who obtained it by the passage of a current from 2000 voltaic cells through two rods of wood carbon, placed end to end, and, after the establishment of the current, slightly separated, producing a light of the most intense brilliancy having the form of an arc; hence the origin of the term *voltaic arc* or *arc-light* by which light, similarly produced, is designated, since it always assumes this form.

It was subsequently produced with 40 Grove or Bunsen cells and rods made of carbon obtained from gas retorts, but remained as a laboratory experiment till brought into practical use, 60 years after its discovery, by the economical generation of electricity by the dynamo.

Electric Candles.—One of the earliest and simplest methods of producing this light for practical use was by the electric candle; that of Jablochkoff, invented in 1872, being the first. It consisted of two carbon rods, each about $8\frac{1}{2}$ inches in length and $\frac{1}{6}$ of an inch in diameter, imbedded in a cylinder composed chiefly of porcelain clay, known as kaolin, at a distance apart of about $\frac{1}{16}$ of an inch, and mounted vertically on a base.

A dynamo current, passed up one rod and down the other, produced the arc light between them above. The kaolin being an insulator, the current was established between the rods by a carbon primer, connecting their upper ends, which was immediately consumed, and the current subsequently maintained by the incandescent carbon vapor. The rods burned slowly downward, consuming the kaolin also, which increased the light by its incandescence. If a candle was accidentally extinguished, a new primer was required to renew the current.

The average duration of a candle was about $1\frac{1}{2}$ hours, but by using a group of 6, with automatic transfer of current, 9 hours continuous light could be obtained.

The upward radiation with downward shadow, and the liability to accidental extinction, led to improvements, among which was the Jamin candle, constructed with 2 carbon rods, inclined toward each other at an angle, and fed downward by clock-work, making contact at the lower extremities for the establishment of the current, and having subsequent automatic separation to form the arc.

The sun lamp of Clerc and Bureau was another similar device, in which the rods were fed downward by gravity, and maintained at the requisite angle and distance apart by a block of marble or magnesia through which they passed. As they did not come into contact, a primer was necessary to establish the current, which was subsequently maintained by the conductivity which the block acquired by the heat, and which served also to prevent accidental extinguishing; the incandescence of the lime in the marble or magnesia increasing the light and modifying its color. The arc was from $\frac{1}{2}$ an inch to 2 inches or more in length, while in the other candles its length was only $\frac{1}{16}$ to $\frac{1}{8}$ of an inch; and the duration of this candle, with one pair of carbons, was about 10 hours.

The Arc Lamp.—But all these devices were comparatively short-lived, and were superseded by the arc lamp, now in general use, which, with various modifications, consists essentially of two carbon rods, as shown in Fig. 93, maintained in a vertical position by automatic feeding devices controlled by the current which produces the light; being at first in contact, for the establishment of the current, but subsequently separated by the superior current strength thus acquired, to the normal distance required to form the arc; further permanent separation being prevented by the increased resistance which the arc acquires by increase of length, which weakens the current, causing the mutual approach of the carbons when the arc becomes abnormally long, or their contact for instantaneous relighting when accidentally extinguished.

The Arc.—The arc thus formed consists of carbon vapor in union with oxygen. Its usual length varies from $\frac{1}{16}$ to $\frac{1}{8}$ of an inch, but for exceptionally strong lights it may be increased to $\frac{3}{4}$ of an inch. Its electric resistance varies from $\frac{1}{2}$ an ohm to 100 ohms, and its illumination from 1000 to 2000 candle-power; its heat intensity being sufficient to volatilize the most refractory substances, not excepting the diamond. Its characteristic form is due to the difference of electric potential between it and the external air, by which it is attracted outward at the center while retaining its attachment to the carbons above and below; the potential difference on its opposite sides being unequal on account of its position being at the side of the central line of the carbons as shown below.

When a direct current is employed, as shown by the

+ and — signs in Fig. 93, a crater is formed in the upper carbon and a point on the lower, and the current producing the arc, following the path of least resistance, passes to the point of the lower carbon from the lowest projection on the irregular rim of this crater. As this

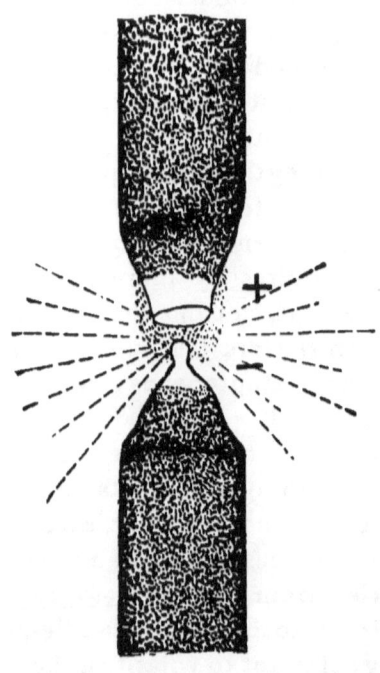

FIG. 93.

projection burns away the arc shifts to the next lowest point and thus travels continuously round the crater above, as if pivoted on the point of the lower carbon.

The Crater and Point.—The formation of the crater is due, in part, to the checking of the current and consequent accumulation of energy above by the high resistance of the arc, causing increased consumption of carbon. The exterior of both carbons is consumed more rapidly than the interior, consumption increasing toward the tips, producing a cone on each, the lower pointed and the upper truncated. There is also, probably, a certain degree of electrolysis, producing excess of oxidation at the anode, or upper carbon, and corresponding diminution at the cathode; carbon vapor forming the electrolytic bath; the intensity of this action at the center, where the vapor is densest, producing the crater and point. In short arcs particles of carbon and fused impurities are deposited on the cathode, forming the *mushroom tip*, shown in Fig. 93, which is burnt off at the base and again renewed as the consumption proceeds.

With the direct current, the positive carbon is consumed about twice as fast as the negative, but with the alternating current the consumption of both is equal, and both become pointed.

The Heat and Light.—The heat is greatest in the carbon vapor, and the light greatest in the incandescent carbon, 65% of it being from the crater, the downward radiation from which is of special importance in the arc light, whose elevation for safety and convenience becomes necessary in consequence of its intense brilliancy and the powerful currents required to produce it.

Establishment of the Current.—The contact of the carbons for the establishment of the current becomes necessary from the fact that a current sufficient to maintain the longest arc cannot pass through an air space of $\frac{1}{10000}$ of an inch, while the momentary condensation of electric energy, and consequent high potential difference produced between the carbons previous to their separation, is sufficient to overcome the high resistance of the air film and cold carbon, and establish the arc, which is then maintained, through the reduced resistance, by the normal current.

The Carbons.—Carbon, originally used by Sir Humphry Davy in the discovery of the electric light, is still found to be the only substance suitable for its successful production; and it is of the highest importance that it should be pure and of homogeneous composition.

Various carbonaceous substances have been employed for the production of the arc-light carbons, as coke, coal, charcoal, lampblack, graphite, and sugar; but petroleum coke, a residuum of the distillation of crude petroleum, has given the most satisfactory results. It is ground and then mixed with some hydrocarbon, as gas-house pitch, and after being thoroughly ground again, is molded in steel molds, heated and condensed by heavy pressure and the infiltration of hydrocarbon, and hardened and purified by repeated baking at various temperatures.

The process involves numerous manipulations and equires great circumspection; the result being the production of carbons of remarkable purity and homogeneousness. They are usually about 12 inches long, and vary in diameter from $\frac{7}{18}$ to $\frac{9}{18}$ of an inch, or more, in proportion to the current and candle-power required. They are beveled for concentration of the current, at the end intended for lighting, and usually copper-plated to within an inch of the point, for increase of conductivity.

Automatic Regulation.—The automatic regulation of the light is accomplished either by a train of clock-work or by a solenoid; both methods being in general use. The first is the oldest and was invented by Foucault, receiving various improvements in its earlier stages by Duboscq, Serrin, and Lontin; further improvements being subsequently added.

In both methods the carbons are attached by sockets and binding-screws to brass rods supported vertically, which, in the first method, are operated by the clockwork by means of electromagnets, through the coils of which the current passes. When the carbons are in contact or too close, the strong current through the magnet coil attracts the armature operating the clockwork and separating them, in opposition to the force of a spring, a weight, or an opposing current, which tends to bring them together; and as the current producing the separation becomes weakened by the increased resistance of the arc a balance between the opposing forces is obtained, by which the arc is maintained at its normal length.

In the solenoid method, used by Siemens, Brush, and others, the upper carbon holder is lifted against the force of gravity by an armature to which it is attached, which moves vertically in the interior of a solenoid coil through which the current passes. As the armature is attracted upward, a clutch attached to it grips the edge of a loose washer, which being tilted grips and lifts the carbon holder which passes through it.

Fig. 94 illustrates this and shows its application to the double carbon lamp, shown in Fig. 95. The clutch on

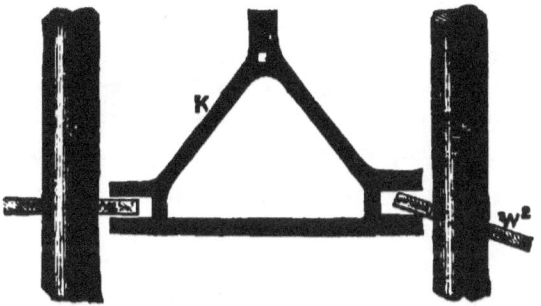

FIG. 94. FIG. 95.

the left being narrower than the one on the right, the left pair of carbons are kept apart by this simple device till the pair on the right are consumed, when the change of resistance instantly brings the left pair into contact, and the light is renewed.

Hefner von Alteneck's Regulator.—The regulator of Hefner von Alteneck, of which Fig. 96 is an ideal illus-

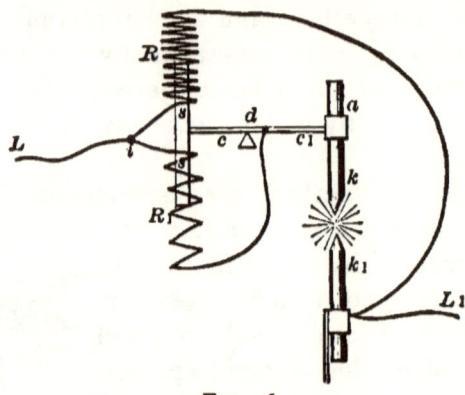

FIG. 96.

tration, has an important application to the solenoid lamp and to arc lighting in general.

The current from L to L_1 divides at i, the main branch going through the low resistance coil R_1 and the lamp, as shown, while a shunt current of about 1% of the entire strength goes through the high resistance coil R and round the lamp. The armature ss is drawn down by the greater magnetism induced by the lower current, separating the carbons and establishing the arc. As the resistance of the arc increases with its length, the potential difference, or E. M. F., between L and L_1 increases, and the strength of the lower current decreases in like proportion. But as the resistance in R remains constant, the strength of its current is increased by the increased E. M. F. in the same ratio as that in R_1 is diminished by the increased resistance, tending to draw the armature *ss* upward by the increased magnetism induced and shorten the arc, which thus becomes adjusted to its normal length and a balance is maintained.

These coils may be arranged in any convenient manner, as by winding in opposite directions, one outside the other; the shunt current thus opposing and partially neutralizing the magnetic effect of the main current, as in the Brush arc lamp.

Series Distribution.—As currents of 10 to 15 amperes are usually required for arc lamps, the series method of distribution is found to be the most economical, and the only practical method; the entire current passing from lamp to lamp through a series often embracing 50 or more, distributed over a large building, or area of a town.

Automatic Cut-Out.—As any variation of resistance in a lamp affects every lamp in the series, regulators, constructed on the principle of Hefner von Alteneck's, are required in the series system; also automatic short-circuiting apparatus for the exclusion of extinguished lamps, without which the extinction of a single lamp would interrupt the current, causing the extinction of every lamp in the series. Such apparatus, in the Brush lamp, consists of an electromagnet wound with two coils, a fine wire coil on a closed circuit connected with the shunt, and a coarse wire coil on an open circuit connected with the magnet's armature. The ordinary shunt current does not induce sufficient magnetism to attract the armature, but the increased current, caused by the extinction of the lamp, is sufficient for this purpose; the attracted armature closing the coarse wire circuit, by which the full current is carried past the extinguished lamp.

The Incandescent Lamp.—In the first attempts to produce the electric light by incandescence exclusively, platinum wire was employed and also iridium, but the superior advantages of carbon were soon demonstrated; consisting in its high electric resistance, 250 times as great as that of platinum, its infusibility at the highest temperature, and its greater illuminating power. But as it is volatilized at high temperatures in the presence of oxygen, its exclusion from the air became necessary, and this was accomplished by inclosure in a glass bulb in which a high vacuum was subsequently produced by a mercury pump. Such are the general principles of construction of the incandescent lamp as we now have it, as illustrated by Fig. 97.

FIG. 97.

The Filament.—The carbon, prepared from a variety of different substances, as bamboo, bass broom, cotton, linen, and silk, consists of filaments bent into any convenient form which will fit in the glass bulb. They are subjected to numerous manipulations to give them the requisite hardness, tenacity, elasticity, homogeneousness, and durability. The principal steps are the forming; carbonizing by baking at a high temperature with exclusion of air; and "*flashing*," which consists in heating the carbonized filaments to incandescence by the electric current or otherwise, in a bath of carbon vapor, the carbon from which is thus deposited on them, forming an even, dense, hard, homogeneous coating. The carbon of some filaments is entirely built up in this way on a base of fine platinum wire. There are also filaments made of hollow tubes for increase of surface.

The average durability of a filament, in the 16 candle-power lamp, is from 600 to 1000 hours; the heating and cooling, molecular action, and general wastage, finally terminating in its rupture, requiring renewal of both filament and containing bulb. Its electric resistance, when heated to incandescence, is about half its cold resistance, ranging from 50 to 200 ohms, according to its length, cross-section, and composition.

Filament and Lamp Attachment.—Each filament, when completed, is attached at both ends, as shown, to platinum terminals sealed into the glass, after which the air is exhausted and the bulb hermetically sealed.

Each bulb is then attached to a socket from which it can be easily removed for replacement; in which is a device, operating with springs, for closing or opening the circuit by turning the insulating handle shown, by which the current is passed through the filament or excluded from it for lighting or extinguishing the lamp.

Position and Current.—The position of this lamp when in use is entirely a matter of convenience, as its illumination seldom exceeds 16 candle-power, and its current $\frac{1}{2}$ to $\frac{3}{4}$ of an ampere. The current may be either direct or alternating according to the system of lighting, each system having numerous distinctive features.

Parallel Distribution.—The large number of lamps required on an incandescent lighting circuit and the small current required for each makes the parallel system of distribution the most economical and practical. This system is illustrated by Fig. 98, in which are represented two heavy copper mains issuing from the dynamo, between which the lamps are mounted on fine wire connections.

These mains may extend to any required distance through a building, or through streets, with branch mains extending into the buildings; but when the direct current is employed, they must be of sufficient size to reduce the resistance to a required minimum. A copper conductor capable of carrying a current sufficient to feed 5000 16 candle-power lamps at a mean distance of 4000 feet from the dynamo would require a cross-section of 12.57 square inches, the size being proportionally reduced as the line branches into parallel circuits, while wire of No. 14 to 16 gauge is large enough for the lamp connections.

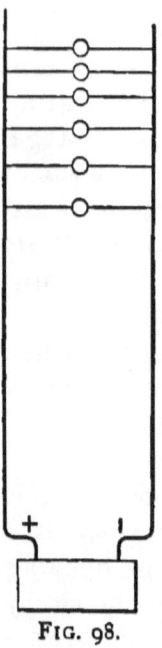

Fig. 98.

If a circuit, like that shown in Fig. 98, have a resistance, including that of the dynamo, of 1 ohm, and each filament a hot resistance of 199 ohms, and the dynamo an effective E. M. F. of 100 volts, then, if a single lamp be lighted, it has $\dfrac{100 \text{ volts}}{200 \text{ ohms}} = \dfrac{1}{2}$ an ampere current. But if two lamps be lighted, the current has two paths instead of one between the mains, which is the same, in effect, as doubling the cross-section of the filament and thus halving its resistance; which gives $\dfrac{199R}{2} + \dfrac{2R}{2} = \dfrac{201R}{2} = 100\tfrac{1}{2}R$; then, if the fraction be neglected, $\dfrac{100E}{100R} = 1C$ for the 2 lamps, ½ an ampere to each, as before.

For any small number of lamps the resistance varies inversely and the entire current directly as the number lighted, and the current per lamp remains practically constant, as shown, being equally divided among the

THE RELATIONS OF ELECTRICITY TO LIGHT. 307

entire number lighted. But as the resistance of the dynamo and circuit remains constant while that of the lamp filaments varies, it is evident that in the lighting of any considerable number of lamps the fraction, neglected above, would make a sensible difference in the ratio of resistance to E. M. F. Suppose that 100 were lighted, then the entire filament resistance would be $\frac{199R}{100} = 1.99R$, and, adding in the 1 ohm constant resistance, we have 2.99 ohms as the entire resistance; hence $\frac{100E}{2.99R} = 33\frac{44}{100}C$, which, divided among the 100 lamps, gives about $\frac{1}{3}$ of an ampere per lamp, instead of $\frac{1}{2}$ an ampere, with only one or two lighted.

There is also a certain amount of current wastage, making an entire current variation of 15% to 20%, which must be provided for in order to maintain constancy of current and illumination. This, in the direct current system, is done by the introduction of resistance coils into the circuit, by which the current can be varied by variation of the resistance, and in the alternating current system by a direct variation of current in the converter.

Hence when the indicator at the station shows a variation of current below or above the normal, by the lighting or extinguishing of any considerable number of lamps, the attendant makes the necessary correction by moving a switch either in the resistance box or converter according to the system of lighting employed.

Multiple Series and Series Multiple.—A number of short series of lamps may take the place of single lamps on a parallel circuit, producing what is termed a "multiple series" installation; or a number of groups with lamps in parallel in each may be placed in series, producing what is termed the "series multiple" installation.

308   *DYNAMIC ELECTRICITY AND MAGNETISM.*

Three-Wire System.—In the Edison three-wire system, illustrated by Fig. 99, two parallel circuits with two dynamos are combined, the dynamos being connected together in series as shown; and a single central main, attached to the short connector which joins them, takes the place of the two interior mains, and equalizes the current through the lamps, in the following manner. When an equal number of lamps is lighted on each circuit, the resistance between the circuits being equally balanced, the entire current flows across through the several pairs of lamps in series between the two external mains. But the lighting of a greater number on one circuit than on the other reduces the resistance and increases the current in that circuit; and this surplus current flows through the central main; in a negative sense if the increase is in the left-hand circuit, but in a positive sense if the increase is in the right-hand circuit.

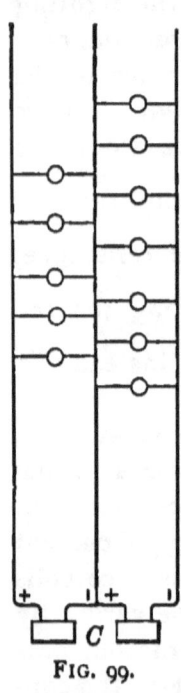

FIG. 99.

Three mains are thus enabled to do the work for which four are usually required. But a further reduction in the required amount of conducting metal results from the fact that this amount is found to vary inversely as the square of the required E. M. F., which being doubled by joining the two dynamos in series, the cross-section of each main should be reduced to ¼ the usual amount, its length remaining the same if there were no change of filament resistance. But the joining of each pair of lamps in series increases the filament resistance to four times the amount of that of each pair joined in parallel, the current traversing twice the filament length with half the cross-section. Hence the ratio of resistance to E. M. F. would be doubled if mains of only $\frac{1}{4}$ the usual size were employed; therefore, to maintain constancy of current, mains of $\frac{1}{2}$ the cross-section and same length would be required; that is, 3 mains, each $\frac{1}{8}$ the usual size, or $\frac{3}{8}$ the usual amount of copper, if each main were required to carry a full current. But, as the central main carries only the required surplus of current, its cross-section can be reduced about $13\frac{1}{2}\%$ below that of the other two.

CHAPTER XII.

THE ELECTRIC TELEGRAPH.

Early History.—The experimental stage of the electric telegraph extends back to the middle of the last century; static electricity having been first employed for the transmission of signals; a plan for alphabetic signaling by this means being described in Scot's Magazine for 1753. Lesage constructed the first electric telegraph, in 1774, at Geneva; in which he employed 24 wires, each connected with a separate pith-ball electroscope, representing a letter of the alphabet. Similar methods were employed later, by Lomond in 1787, and Ronalds in 1816. Reusser, in 1774, suggested the illumination of letters made with metal spangles on glass plates, as an improvement on the pith-ball method of Lesage.

Sömmering, in 1808, first employed voltaic electricity for telegraphing, using 35 water voltameters, each connected with separate wires and giving separate signals: and similar methods were subsequently tried by Coxe, Smith, Bain, and others.

Ampère, in 1820, proposed to employ 24 galvanometer needles, each connected with a separate wire. Schilling, in 1832, and Weber and Gauss, in 1833, employed a single needle, indicating alphabetic signals by right and left deflections. Steinhill subsequently developed this system, employing two needles, one for the positive and the other for the negative current, both deflected in the same direction; alphabetic signals being given by bells struck by the needles, and also by dots made with ink on a moving strip of paper, as well as by observation of the movements by the eye.

Steinhill, while constructing a telegraph line at Munich, in 1838, made the very important discovery that the current could be carried by a single wire, and the earth employed for the return circuit by making connection with it at the terminals of the line; from which he inferred that the earth took the place of the return wire as a conductor; but subsequent experiments seem to prove that the earth, in this case, is to be regarded as an electric reservoir, giving and receiving electric energy, rather than as a conductor.

Cook and Wheatstone, in 1837, introduced the *needle* telegraph, as it was designated, into England, and constructed, on the London and Birmingham Railway, the first line ever employed for commercial use. It consisted of five underground wires connected with five separate needles; a system which they subsequently modified, employing two wires connected with two needles in one method, and a single wire and needle in another method. The signal for the transmission of a message was given by a bell rung by an electromagnet.

In 1831 Henry transmitted signals by sounds produced by the movements of the armature of an electromagnet; and Morse, in 1835, invented a telegraph operated in a similar manner, in which alphabetical signals, consisting of lines and dots, were made on a moving strip of paper, first by a pencil, but subsequently by a steel point which embossed them on a grooved roller over which the paper was moved by clock-work operated by a weight.

Morse constructed the first commercial telegraph line in the United States, between Washington and Baltimore, and sent the first message over it May 27, 1844.

This line was mounted on wooden poles and consisted of two iron wires, the practicability of employing the earth for the return circuit being then imperfectly understood.

The American Morse Code.—The original Morse code for letters, numerals, and punctuation, now employed in the United States and Canada, is as follows:

A	B	C	D	E	F	G	H	I
·—	—···	·· ·	—··	·	·—·	——·	····	··

J	K	L	M	N	O	P
—·—·	—·—	———	——	—·	· ·	·····

Q	R	S	T	U	V	W	X
··—·	· ··	···	—	··—	···—	·——	·—··

Y	Z	&	Period	Semicolon	Comma
·· ··	··· ·	· ·—·	··——··	·—·—·	·—·—·

Exclamation	Interrogation	Paragraph	Parenthesis	Italics
———·	——·—··	——— ··	·——·—··	———·—··

1	2	3	4	5	6
·——·	··—··	···—·	····—	———	······

7	8	9	0
——··	—····	—··—·	———

It will be noticed that this code consists of long dashes, short dashes, dots, and spaces; L for instance being indicated by a long dash, T by a short dash, R by three dots with space between the first and second, C by three dots with space between the second and third. Hence the number and relative positions of these four elements constitute the distinction between the different characters; the spaces having equal significance with the dashes and dots.

The International Morse Code.—The Morse code has been found so well adapted to telegraphing, that it has superseded all others for this purpose, and come into general use throughout the world. But on its introduction into Europe some changes were necessary to

THE ELECTRIC TELEGRAPH. 313

adapt it to the various languages, and also to remedy defects which had been developed by its practical use in America. This led to the adoption, by a telegraphic convention assembled at Vienna in 1851, of the international Morse code, now employed in all countries except the United States and Canada. In this code long spaces between the elements of a letter are eliminated, as they are liable to be misunderstood for the spaces between letters; each numeral is represented by five elements, and each punctuation mark by six. The differences between this code and the American are as follows:

C	F	J	L	O
−·−·	··−·	·−−−	·−··	−−−

P	Q	R	X	Y
·−−·	−−·−	·−·	−··−	−·−−

Z	Ch	Ä	Ö	Ü
−−··	−−−−	·−·−	−−−·	··−−

É	Ñ	Period	Comma
··−··	−−·−−	·−·−·−	−−··−−

Exclamation	Interrogation	Apostrophe	Hyphen
−−··−−	··−−··	·−−−−·	−····−

Parenthesis	1	2	3
−·−−·−	·−−−−	··−−−	···−−

5	6	7	8	9
·····	−····	−−···	−−−··	−−−−·

0
−−−−−

As it was soon found that messages could be read more easily and rapidly by the click of the instrument than by the record on the paper, the dots, dashes, and spaces came to indicate sounds and pauses, and the registering instruments were replaced by sounders in all the principal offices.

Simple Line Equipment.—The principal apparatus required for the equipment of a simple Morse telegraph line are a *battery, signal key, sounder* or *register*, and *relay;* all of which must be duplicated at each end of the line; the duplication of the battery, on such a line, being desirable though not always strictly necessary. A *lightning arrester*, *ground switch*, and *cut-out* are also required.

The Battery.—The principal requirements of the battery are strength and constancy, and any good battery fulfilling these conditions can be employed. The gravity battery is one of the best and cheapest, and hence is extensively used for this purpose. It requires comparatively little care, is free from noxious fumes, and not liable to polarize.

The Key.—The key, one form of which is shown in Fig. 100, is a lever of steel or brass, so mounted as to

FIG. 100

have a vertical movement, limited by two set-screws which can be adjusted to any required range of motion; the upward movement being produced by a spring connected with one of the screws, and the downward by pressure on the hard-rubber knob at the left, which closes the circuit by bringing a little projection underneath into contact with an anvil attached to the left hand bolt; the points of contact being faced with platinum to prevent fusion by the extra current at break.

This bolt and anvil are insulated from the supporting frame, while the bolt at the right is connected with it, and both are connected with the terminals of the electric circuit. When the key is not in use the circuit is closed by a lever pushed under a metal projection attached to the anvil.

The Register.—A simple form of the embossing register is shown in Fig. 101. The armature of an electromagnet M is attached to the bent lever L, pivoted at d so

FIG. 101.

as to have a vertical movement limited by the adjustable screw m and stop underneath. A steel point p, attached by an adjustable screw to the bent end of the lever, makes contact in a little groove with the roller r, and embosses the message on a strip of paper carried between the rollers, which are operated by clock-work impelled by a weight attached to a cord wound on the drum W, and controlled by the brake a. The electromagnet is connected with the line by binding-posts, one of which is shown at s, and when the current, transmitted from the distant station, attracts the armature, the lever L is drawn down against the force of the spiral spring n, bringing the point into contact with the paper, and registering the message as described.

Double embossing registers, operated on the same principle, are now in common use, by which two separate messages can be registered in parallel lines on the same strip of paper, or one message only, on a single line, as required. Inking registers, both double and single, are also in common use, and are generally preferred to the embossing instruments. The clock-work, in all the new registers, is operated by a spring.

The Sounder.—One of the best known forms of the sounder is shown in Fig. 102. A bent lever, having a

Fig. 102.

vertical and a horizontal arm, is pivoted on an arched support as shown, the vertical arm being concealed by the support. The horizontal arm has a vertical movement between the poles of an electromagnet, limited by the adjustable set-screws shown above on the left; and is held in contact with the upper screw, when the circuit is open, by a retractile spring connecting the lower end of the vertical arm with an adjustable screw which passes through the supporting post on the left.

The instrument is connected with the line by the binding-posts on the right, and when a current is sent through the coils of the magnet the attraction of the armature brings down the lever, the point of the screw striking with a sharp click on the curved brass sounding piece. When the current ceases the spring brings the lever up with a light click against the screw above; and by means of these two clicks, signals indicating the dots, dashes, and spaces are distinguished. The sharp click indicates the beginning of a dot or dash, and the light click its termination; a pause following a sharp click indicates a dash, and a pause following a light click indicates a space.

The screws can be adjusted to any required range of motion, both in the sounder and register; and the armature, in both instruments, is kept out of contact with the magnet poles, to prevent magnetic adhesion.

The Relay.—On short, well-insulated lines, not exceeding 20 or 30 miles in length, the sounder, or register, if its resistance is not too high, can be operated by the line current; but, on longer lines, resistance and imperfect insulation usually weaken this current too much for direct action; but by the aid of a *relay* it can perform this work indirectly through the agency of a local battery current.

A common form of the relay is shown in Fig. 103. An electromagnet supported in a horizontal position by an adjustable screw, on the right, and a curved standard, on the left, has its armature attached to a vertical lever, pivoted below, but having a horizontal movement above limited by two screws by which its range of motion can be adjusted. A retractile spring holds it against the point of the left hand screw when the circuit is open, while a weaker spring below tends to force it in the opposite direction; the tension of the upper spring being capable of adjustment as shown.

The two binding-posts on the left are connected respectively with the curved standard and lever support, and, exteriorly, with a local battery which embraces in

FIG. 103.

its circuit the sounder or register; and the electromagnet is connected with the line by the two binding-posts on the right. Hence, when the line current passes through the magnet coils, the armature is attracted and the local circuit closed by contact between the platinum points attached to the lever and right hand screw, and the sounder or register operated by the local current.

The distance between the magnet and its armature can be adjusted by the supporting screw on the right, which moves the magnet through the openings in the curved support on the left. Hence the adjustment by this means and that of the retractile spring and upper screws can be adapted to any current which may be sent over the line.

Cut-Out, Ground Switch, and Lightning Arrester.—These three instruments, of which there are several different forms, are employed separately, or may all be combined in one. A common form of the latter kind is shown in Fig. 104. Three brass plates, with binding-posts attached, are

FIG. 104.

mounted on an insulating block, the central plate having a row of points on each side. This block is attached to the wall in any convenient position, the end plates connected with the terminals of the line, and the office instruments placed in circuit between them, and the central plate connected with the earth.

When a brass plug is inserted between the end plates, as shown, the office instruments are cut out of the circuit, but when the plug is removed the current must evidently pass through the instruments. If, under the latter arrangement, lightning should strike anywhere on the line, its high potential would cause its current to pass to the earth by way of the points and central plate, instead of taking the longer route through the instruments.

When the line connections, at a way station, are interrupted, the direction in which the interruption has occurred may be ascertained, and current from the opposite direction obtained by making connection with the earth. This is done by inserting the plug between the central plate and one of the end plates. If the interruption were on the right and the plug should happen to be first inserted on the left, no current would be obtained; if then the plug were inserted on the right, the instruments would be placed in circuit between the earth and the uninterrupted connection on the left and current obtained; and in like manner connection could be established on the right, if the interruption were found to be on the left.

The cut-out should always be closed during a thunderstorm or the absence of the operator, to prevent accidents to the instruments.

Line Construction.—The ordinary telegraph line is constructed with No. 6–7 iron wire, B. and S. gauge, but for short lines No. 8–9 wire can be used. It is coated with zinc to prevent oxidation, and mounted on wooden poles, provided with supporting cross-arms to which it is attached by glass insulators; a large number of parallel wires, arranged in tiers, being often mounted on the same poles.

In cities where air lines are prohibited, the wires, coated with insulating material and combined in cables, are laid underground; the cables being inclosed in lead pipes, or otherwise protected against moisture and abrasion, and often placed in conduits so as to be accessible without disturbance of the pavement.

Joints are made, where required, by twisting the wires firmly together, and then soldering them to insure perfect electric connection and prevent its interruption by oxidation; but electric welding, now coming into use, is preferable for this purpose. Where wires pass through walls, for office connections, they require to be well insulated by hard-rubber tubes.

THE ELECTRIC TELEGRAPH.

Station Arrangement.—The arrangement of the instruments and connections of a terminal station are shown in Fig. 105. The line is connected with the earth at *G*, the wire being soldered to a mass of buried metal for good connection, or to water or gas pipes, where they are available for this purpose. It is connected with the main battery shown at *E*, which has its positive pole connected to the earth and its negative to the line; this

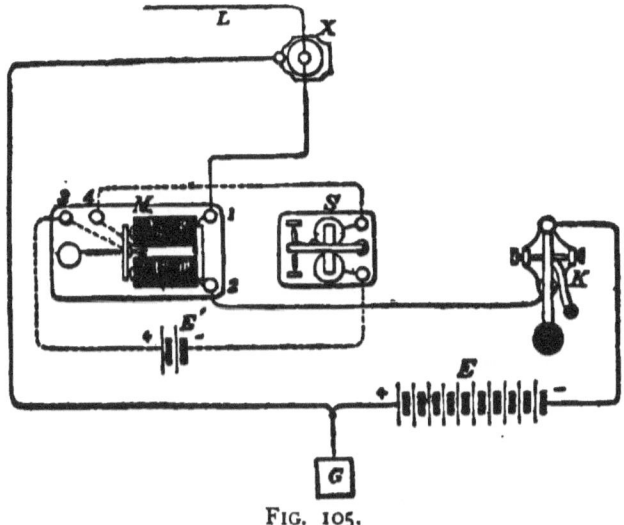

FIG. 105.

arrangement being reversed at the opposite end of the line; hence the line current, at this station, always flows *towards* the battery, and at the opposite terminal station, *from* the battery, while the earth current, at each station, flows in the opposite direction. The instruments and battery may be cut out, when necessary, by a switch or plug connection, at *X*, with the ground wire shown at the left.

When a message is being received, the circuit is closed through the key *K*, and the line current, entering through the lightning arrester at *X*, traverses the relay

M by the binding-posts 1 and 2, and goes through K and battery E to the earth at G. The circuit of the local battery E', being closed by the relay, its current passes from the positive pole, by the binding-posts 3 and 4 and platinum points, through the sounder S and thence to the negative pole of E'.

When a message is being sent, the circuit is opened at K, and the instruments at the distant station, traversed by the *outflowing* current, respond to the manipulations of the key at the home station, where the instruments are traversed by the *inflowing* current as before. The same conditions, with reversal in the direction of the current, occur at the opposite end of the line.

It is important that the instruments should always be in circuit during business hours, as a station is liable to be called at any moment; and their constant click is a notice to the operator that his connections are right and the line in working order. All the messages and station calls are therefore heard at every station on the line, but responded to only at the station called.

The arrangement of a way station is the same as that of a terminal station except that only the local battery is required; the current from the main batteries entering by one branch of the line, traversing the relay and key and operating the sounder by the local battery, and leaving by the other branch of the line. Messages can therefore be received or sent in either direction, the current being positive to all the stations on one side and negative to all those on the other.

The current is derived from both terminal batteries, which may be regarded as a single battery, with the instruments interposed between its poles in one branch, and the earth in the other. Hence if either battery is

THE ELECTRIC TELEGRAPH. 323

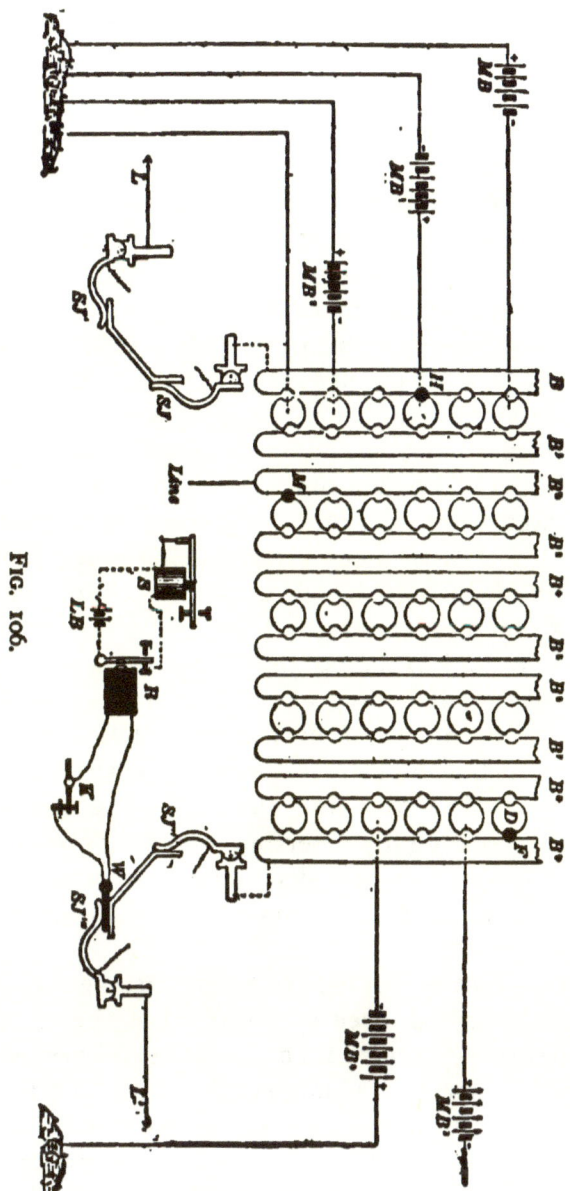

FIG. 106.

cut out by a ground switch, either at a terminal or way station, the current is proportionally reduced.

Switch Board.—When a number of different lines have connections through the same office a switch-board, like that represented by Fig. 106, is required. The vertical brass bars represent line connections and the rows of brass disks, battery connections. The disks in each horizontal row are electrically connected together at the back, but insulated from the other rows; and each row, except the lowest, connected with a separate battery, the lowest being connected with the earth.

By the apparatus known as a spring-jack the instruments may be connected with the line as shown at SJ'''; a wedge W having brass plates on its opposite sides, insulated from each other, but connected with the terminals of the instrument circuit, being inserted between a pair of springs, which close the circuit again when the wedge is withdrawn.

The bars B and B^2 being thus connected with the line, the insertion of a plug at H puts battery MB' in connection with the line L by the vertical bar B; and, in like manner, battery MB is connected with L' by the plug F and bar B^2. Any line may be connected with the earth for testing by inserting a plug between its bar and the lowest row of disks. Thus the insertion of a plug at M gives the line connected with B^2 an earth connection.

Repeaters.—As the distance to which messages can be transmitted is limited, even with the aid of the relay, it becomes necessary to have them automatically repeated by a special apparatus which employs local batteries. By this means stations four or five thousand miles apart can hold communication with almost the same facility as those on short lines. Press despatches can also, in this way, be received simultaneously by all the principal stations on a line.

Before the introduction of this method the message had to be repeated by the operator, involving great delay and liability to errors.

Two sets of instruments are required for repeating, each connected with a separate branch of the line, and including a relay, sounder, and also two or more local batteries to each set.

The Button Repeater.—The *button repeater*, invented by Wood in 1846, and still employed to a limited extent, was one of the first in use. It consists, as improved, of a button switch placed between the two sets of instruments, having double contacts on each side, one pair of which may be closed when the other pair is opened, or both pairs opened as required. Each set of instruments is in circuit between each pair of contacts, and when either pair is closed the two branches of the line are connected through both sets of instruments, and each connected also with a separate main local battery.

When a message is to be repeated, the operator at the repeating station, on being notified, closes the switch contacts through the branch of the line wishing to transmit, giving it a closed connection to the earth through the main local battery of the repeater, with which it is connected; which enables the operator at the sending station to repeat into the opposite branch through the sounder connected with his branch, which also acts as a transmitter. When a message from the opposite direction is to be repeated, the connections are reversed by reversal of the switch.

In this repeater both sets of instruments respond to the manipulations of the sending operator's key, but the closed switch contacts prevent any break in the through connections. When, however, the switch is

326 *DYNAMIC ELECTRICITY AND MAGNETISM.*

disconnected from the contacts on both sides, the through connection being opened, each branch of the line becomes independent of the other, and terminates in its main local battery at the repeating station. Messages can then be sent or received by each set of instruments, separately, by connecting them with local keys, or through connections, independent of the repeating apparatus, made with a single set.

The Milliken Repeater.—The annoyance and delay occasioned by the absence of operators at repeating

FIG. 107.

stations led to the invention of repeaters which can be reversed from distant stations by the current, automatically. The Milliken repeater, shown in Fig. 107, is one of the best known of these. It consists of the ordinary relay, shown at the right, by which the sounder is operated in the usual manner, and an extra magnet, mounted at the left, whose armature is arranged to close the circuit automatically, and keep it closed, through one branch of the line, by the aid of connected apparatus, during the repeating of a message by a similar companion instrument connected with the other branch. The upper retractile spring has greater tension than the lower, so that when the attraction of the extra magnet ceases, its armature is pulled back, bringing an insulated stop against the armature of the relay, as shown, and closing the local circuit through the connected sounder.

Repeater Connections.—The arrangement and connections of the Milliken repeater, at a station, are shown in Fig. 108. T and T' are sounders, used also as transmitters, and connected respectively, at Y and Y', with the extra magnets ExM and ExM' of the opposite repeaters through the circuits of the extra local batteries XL and XL'. The levers of T and T' are furnished with continuity-preserving springs s and s', insulated from them, which make contact with stops above, when the levers are attracted, and close the circuits of the main local batteries MB and MB' respectively; the lever projections, at x and x', limiting the upward movement of the springs, when the attraction ceases, and breaking the contact. In this way each line circuit is closed by a spring through its main local battery before the closing of the circuit at the opposite end of the lever through the extra local battery, and remains closed till after the latter circuit is opened. Thus when the lever of T is attracted, the circuit of MB is closed at s before that of XL is closed at Y, and remains closed till after the contact at Y is again opened. The dark space at the mounting of these springs shows hard-rubber in-

328 DYNAMIC ELECTRICITY AND MAGNETISM.

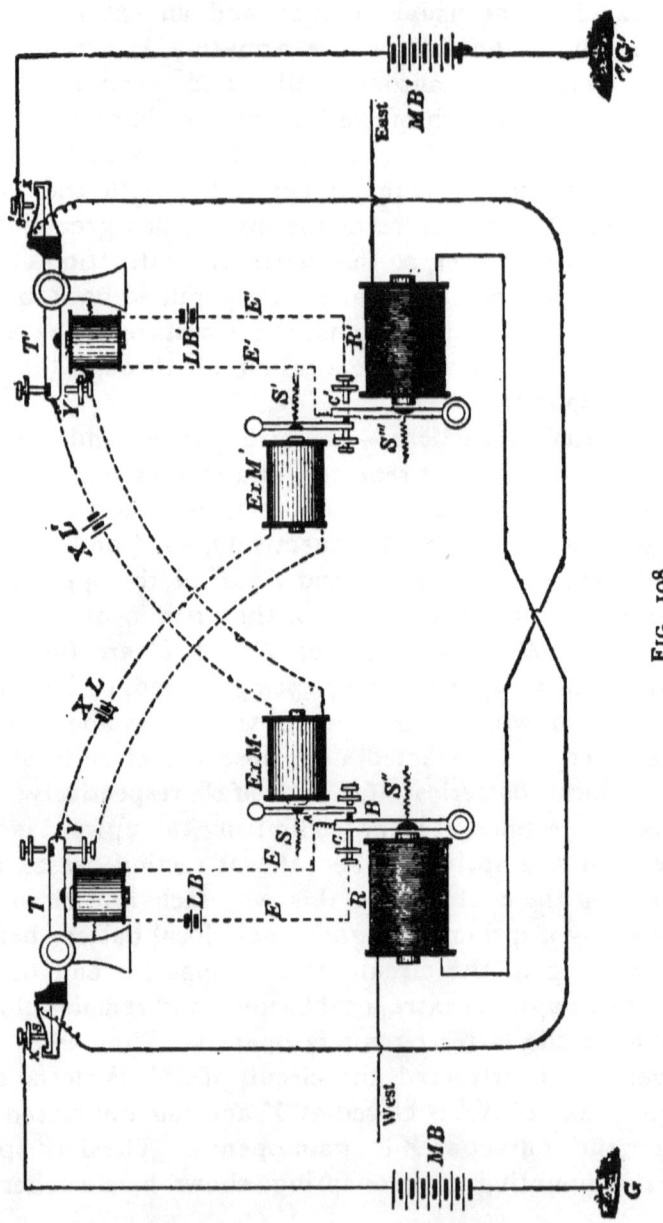

Fig. 108.

sulation, which is similarly indicated at various other points.

The main local battery MB is connected with the eastern line through relay R', and MB' with the western line through R; each battery being connected with the earth as shown.

When a message from the east is to be repeated to the west, the eastern operator opens the circuit through his key, and the operator at the western terminal station, finding this to be the case, closes the circuit through his key; hence there is current in the western line through relay R, but none in the eastern line through R'; armature B is therefore attracted, closing the local circuit EE at C; and transmitter T, being attracted, first closes the eastern line at s, through battery MB, and then the circuit of XL at Y. The magnet ExM' therefore attracts its armature, allowing the spring S''' to open the local circuit $E'E'$ at C': this releases T', opens the circuit of XL' at Y', and then the western line at s', breaking connection with battery MB'. This break stops the current on the western line, demagnetizing relay R; but R's armature is still held closed by the superior force of the spring S over that of S''; hence the connections on the left remain closed while the instruments on the right respond to the manipulations of the eastern operator's key, enabling him to repeat into the western line.

In a similar manner the western operator, by opening the circuit through his key while that through the key at the eastern terminal station is closed, can produce automatic reversal of the connections at the repeating station, and repeat into the eastern line.

Duplex Telegraphy.—The simultaneous transmission of messages in opposite directions on the same wire occupied the attention of various inventors from 1852 to 1872. The first suggestion of the practicability of this method was made by Moses Farmer, an American, in 1852, and the first invention of the kind by Gintl, an Austrian, in 1853. Gintl's invention not proving sufficiently practical, improvements on it were made, in 1854, by Frischen, Siemens, and Halske, and as the result of their labors the duplex system was first put in successful, practical operation, in 1855, between Munich and Vienna, and subsequently, in the same year, between Vienna and Trieste.

Preece, Nystrom, Maron, and other European inventors made various valuable contributions to the duplex system between 1855 and 1863, but it was almost unknown in America till 1868, when Stearns began a series of experiments based on the European methods which resulted, in 1872, in the practical adoption of his system in the United States.

The Stearns Duplex.—The construction and operation of this system is shown in Fig. 109, in which the connections at terminal stations on the right and left are represented. R and R' are differential relays connected with sounders not shown, in each of which the two bobbins are each oppositely wound, as shown, so that currents of equal strength in each would neutralize each other's magnetic effect on the cores, while a current in either branch of the circuit alone, or of greater strength in one branch than in the other, would magnetize the cores. Rh and Rh' are rheostats whose resistance is so adjusted that the resistance from the central point a or a' of either relay through the rheostat to the earth is just equal to the resistance in the opposite direction through the line. T and T' are transmitters (not sounders) operated by the keys and small local batteries shown, and whose levers A and A' have the continuity preserving springs, z and z', already described. MB and MB' are the main batteries, connected with the line through

THE ELECTRIC TELEGRAPH.

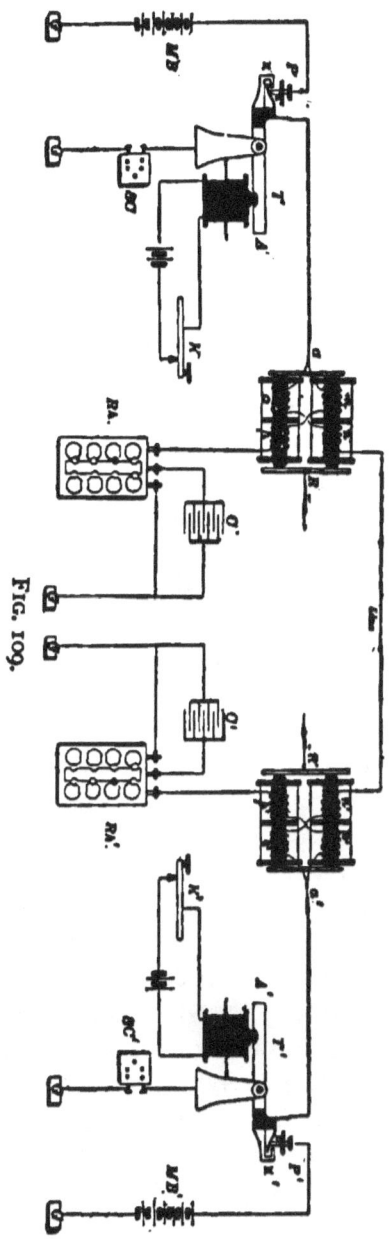

FIG. 109.

z and z', and also connected with the earth. SC and SC' are resistance-coils, also connected with the line and earth, whose resistance is made equal to that of the batteries MB and MB' respectively, so that the line resistance to the earth shall be the same whether the connection is through the coil or through the battery. C and C' are condensers adjusted to absorb a charge equal to the static charge absorbed by the line, and return an opposing current equal to the return current produced by that charge, and thus neutralize it and prevent a false signal.

When a message is to be sent from the station on the left, the depression of key K closes the local circuit through the magnet of transmitter T, and the consequent attraction of lever A closes the circuit of battery MB, through spring z, and opens the ground circuit through SC at x by the depression of the spring as shown.

A current from MB therefore flows through both branches of relay R, and the resistance being the same in each, divides equally at a, one half going to the line through o, n, and the other half to the ground through m, p, and rheostat Rh, except the portion absorbed by condenser C. Hence, no magnetic effect being produced in R, its armature is not attracted; but the line current entering relay R' passes only through branch n', o', hence the core of R' is magnetized and its armature attracted, producing a down click in the connected sounder.

Now let key K' be also depressed, closing the circuit of MB' through spring z', and opening the ground circuit through SC' at x'; a current then flows to MB' from MB through branch n', o', of R'; this doubles the current through n', o', and hence the armature of R' is still attracted and kept closed. But it also doubles the current through branch o, n, of relay R, magnetizing its

THE ELECTRIC TELEGRAPH. 333

core, and hence attracting its armature and producing a down click in the connected sounder.

Now let key K be opened, and the circuit of battery MB through spring z being thus opened, its current ceases; but the ground circuit through SC to battery MB', being closed through x before the circuit through z is opened, R is still magnetized by MB'''s current, and hence its armature remains closed. But MB's current being removed from n', o', of relay R', and the current of MB' flowing equally through m', p', and o', n', R' is demagnetized and its armature released, producing an up click in its connected sounder.

Now let key K' be opened also, and the current of MB' ceases, demagnetizing relay R, whose armature being thus released, an up click is produced in its connected sounder also.

Each sounder therefore responds only to the manipulations of the key at the opposite station and is unaffected by the manipulations of the home key, and hence messages are transmitted simultaneously in opposite directions over the same wire with the same facility as in single transmission. But it is evidently essential to the successful operation of this system that the adjustment of the resistances in both sets of instruments shall be carefully maintained, so as to produce currents of equal strength in both branches of each relay. Each of these currents has the same strength as if the current had not been divided, the E. M. F. and resistance remaining the same in each branch.

The Polar Duplex.—The polar duplex system, as improved, originated from the various inventions which culminated in the invention of the improved quadruplex system. Its two principal instruments are the *pole-changer* and *polarized differential relay*, the latter the invention of Dr. Siemens.

334 DYNAMIC ELECTRICITY AND MAGNETISM.

The Pole-Changer.—The pole-changer, shown in Fig. 110, is constructed with a lever operated by an electro-magnet in opposition to a retractile spring, by means of a small local battery and connected key. The end l of

FIG. 110.

this lever projects, without contact, through an opening in a metal disk D, and has a free vertical movement between two curved, metal springs s and s', connected with opposite poles of the main battery; s being, for convenience, supposed to be connected with the positive pole, and s' with the negative. These springs are attached to the disk by insulated connections, and in proximity with them are two metal blocks b and b' attached to the disk by uninsulated connections, and having adjustable stops with which each spring can make contact alternately. The disk is connected with the line wire, giving these blocks a line connection, and the lever, which is insulated from it, is connected with the earth wire.

When the key is depressed and the lever attracted, it makes contact above with spring s', lifting it and breaking its contact with block b' and also its own contact with s, which therefore springs up into contact with block b. Hence the positive pole of the battery being now connected with the line through s, current flows from it to the line through s and b, and thence to the earth through the apparatus at the distant station; and the negative pole being connected with the earth through l and s', current flows to it from the earth through l and s', and through the battery to the positive pole, completing the circuit.

But when the key is opened, the attraction ceases, and the lever being pulled down by the retractile spring, makes contact below with s, depressing it and breaking its contact with b and also its own contact with s', which therefore springs down into contact with b' Hence, the negative pole of the battery being now connected with the line, the direction of the current is reversed, and it now flows from the positive pole through s and l to the earth, and from the earth at the distant station through the apparatus to the line, and from the line through b' and s' to the negative pole, and through the battery to the positive pole, completing the circuit.

336   *DYNAMIC ELECTRICITY AND MAGNETISM.*

Hence, the polar connections being reversed with each opposite manipulation of the key, the direction of the line and earth currents is correspondingly reversed,

FIG. 111.

while the direction of the current through the battery remains, of course, unaltered.

The Polarized Relay.—The polarized relay, shown in Fig. 111, is constructed with a curved steel magnet, whose north pole, N, we may, for convenience, suppose to be at the upper end, and its south pole, S, at the lower end, as marked. On the south pole is mounted an electromagnet, attached by its soft-iron yoke Y, Y, having oppositely wound coils, M and M, between whose cores the soft-iron armature a, hinged to the north pole, has a free horizontal movement, limited by the stops attached to c and c, with which its projecting brass tongue b makes contact alternately.

This armature derives north polarity from its attachment to the north pole of the permanent magnet, while the cores of the electromagnet derive south polarity from their connection through the yoke with the south pole. Hence, when there is no current in the electromagnet, its poles exert equal and opposite attraction on the armature; and the same is true when currents of equal strength flow through both coils, neutralizing each each other's magnetic effect. But when current flows through only one coil, or is stronger in one than in the other, the permanent magnetism of the cores is overcome, and each acquires polarity in accordance with the direction of the current. Hence the armature, having north polarity, is attracted by the south pole and repelled by the north, and therefore vibrates between the poles in response to the changes in the direction of the current, produced by the pole-changer through the manipulations of the key; alternately closing and opening the local circuit of the connected sounder.

The various connections are made through three pairs of binding-posts on the left; and the positions of the electromagnet, armature, and contacts, and the range of the armature's movement, are adjustable by screws, as shown.

It is practically impossible to prevent the armature from being attracted against one of the stops when there is no current in the magnet coils, notwithstanding the equal and opposite attraction; but such contact is then of no consequence, and, when the instrument is in operation, the armature is always so attracted, normally, by the electromagnetism.

Operation of the Polar Duplex.—The connections and operation of the polar duplex, at opposite stations, are shown in Fig. 112. The line circuit, and the earth, or "artificial," circuit have equal resistances, as in the Stearns duplex.

Let battery B have its positive pole to the line and its negative to the earth, and the corresponding polar connections in battery B' be reversed. The positive current from battery B then divides at S into currents of equal strength, traversing the oppositely wound coils a and b of relay R without magnetic effect; but the current through a is doubled by the negative current of battery B', which flows in the same direction, producing a south pole in the core of a and a north pole in that of b, and hence armature M, having north polarity, is attracted to the pole of a and repelled from the pole of b. The negative current of B' flows in through coils a' and b' of relay R' equally, producing no magnetic effect, but the current through coil a' is doubled by the inflowing positive current from B, producing a south pole in the core of a' and a north pole in that of b'; hence armature L' is attracted to pole of a' and repelled from pole of b'.

Supposing the above conditions to exist when messages are about to be sent in opposite directions simultaneously; let the connections through pole-changer A' be reversed by depression of the key, putting the positive pole of B' to the line; the two batteries being now in opposition positively, there is no current in the line, but there is still a current to the negative pole of each bat-

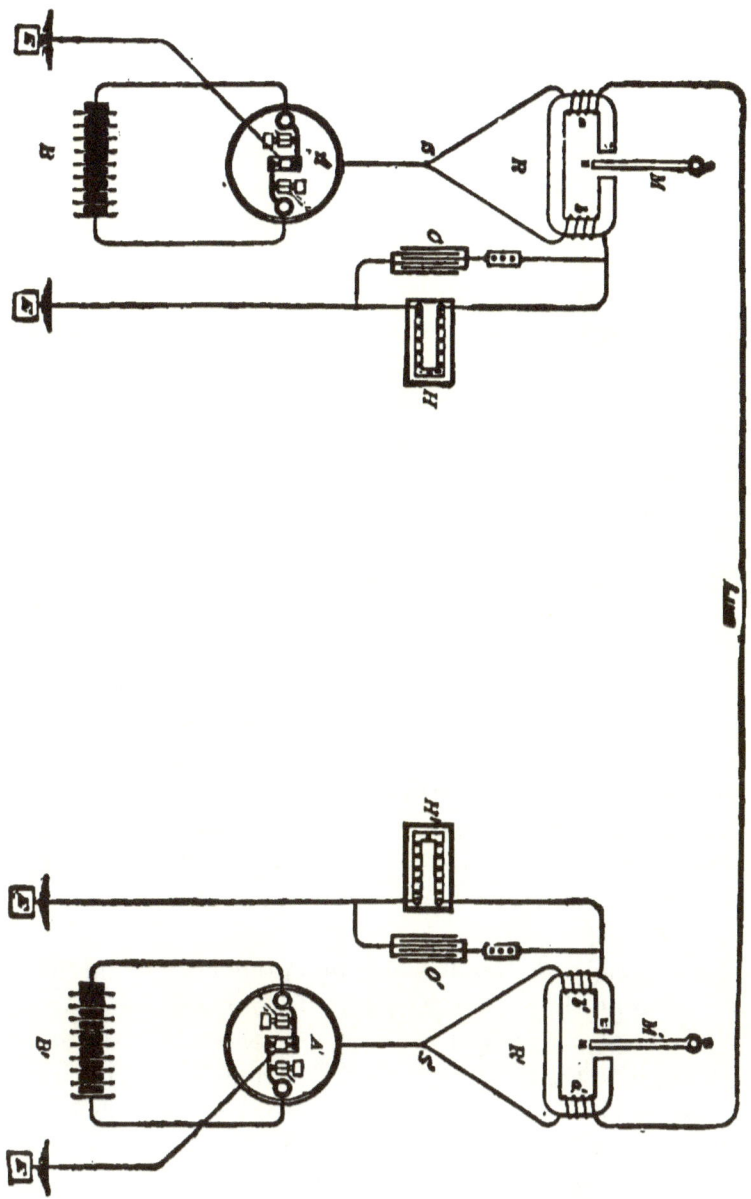

tery from the earth. This current flows to the negative pole of H from right to left through H, coil b, and pole changer A; and hence reverses the polarity of relay R', attracting M from pole of a to pole of b; while that to the negative pole of H' flows as before, from left to right, and hence produces no change of polarity in R'', whose armature M' therefore remains attracted to pole of a' as before.

Now let the polar connections of H' be again reversed, and the former connections being restored, the current through b of relay R' is reversed, and M attracted from pole of b to pole of a, while M' is still attracted to pole of a' as before. Hence relay R' responds to the changes of polarity produced by pole changer A', while relay R'' is unaffected by them.

If, under either of the above conditions, the polar connections of H be reversed, corresponding effects are produced in relay R'', while relay R' remains unaffected.

Sounders, connected with each relay, are operated by local batteries in the usual manner. There is also a small rheostat connected with each condenser, as shown, to regulate its action by retarding its discharge, otherwise liable to be premature on long lines.

Quadruplex Telegraphy. Dr. J. B. Stark of Vienna, in 1855, invented the first experimental method of *diplex*, or double, transmission, in the same direction, on a single wire.

Dr. Bosscha of Leyden also invented a diplex method, at the same time; and Kramer, during the same year, made an improvement in Stark's method, and Maron of Berlin, in 1863, improved Bosscha's method. But none of these methods came into practical use, though beneficial in opening the way for future inventors, and no further progress in diplex transmission was made till after the practical adoption of the Stearns duplex system in 1872.

Dr. Stark, in describing his system, in October 1855, was the first to show that diplex transmission could be combined with duplex, so that four messages could be transmitted simultaneously in opposite directions on the same wire; but he did not think such a system could become practical. Dr. Bosscha also showed the possibility of simultaneous quadruple transmission, in describing his method.

In 1873, Oliver Heaviside, an English electrician, showed that not only was simultaneous quadruple transmission practical, but also simultaneous multiple transmission, to an indefinite extent; so that, theoretically, any required number of messages could be transmitted simultaneously in opposite directions. Multiple transmission, to this extent, has not yet been practically realized; but it has been experimentally shown to be possible, to the extent of eight or even sixteen messages, while quadruple transmission has long since become a practical success.

In 1874 Edison, in connection with Prescott, invented the first quadruplex telegraph. It was constructed on the principle of the Wheatstone bridge, and was operated between New York and Boston by the Western Union Telegraph Company. In 1875 and 1876 Gerritt Smith made important improvements in this system, and the quadruplex, as now constructed, is the result of the combined labors of Edison, Smith, and Prescott, in improving, combining, and applying apparatus previously invented by others.

Construction and Operation of the Quadruplex.—The Edison quadruplex system, as now operated by the Western Union Telegraph Company, is a combination of the Stearns duplex and polar duplex, in such a manner that the two methods can be operated simultaneously, on the same line, without interference. Fig. 113 shows the construction and connections at two terminal stations,

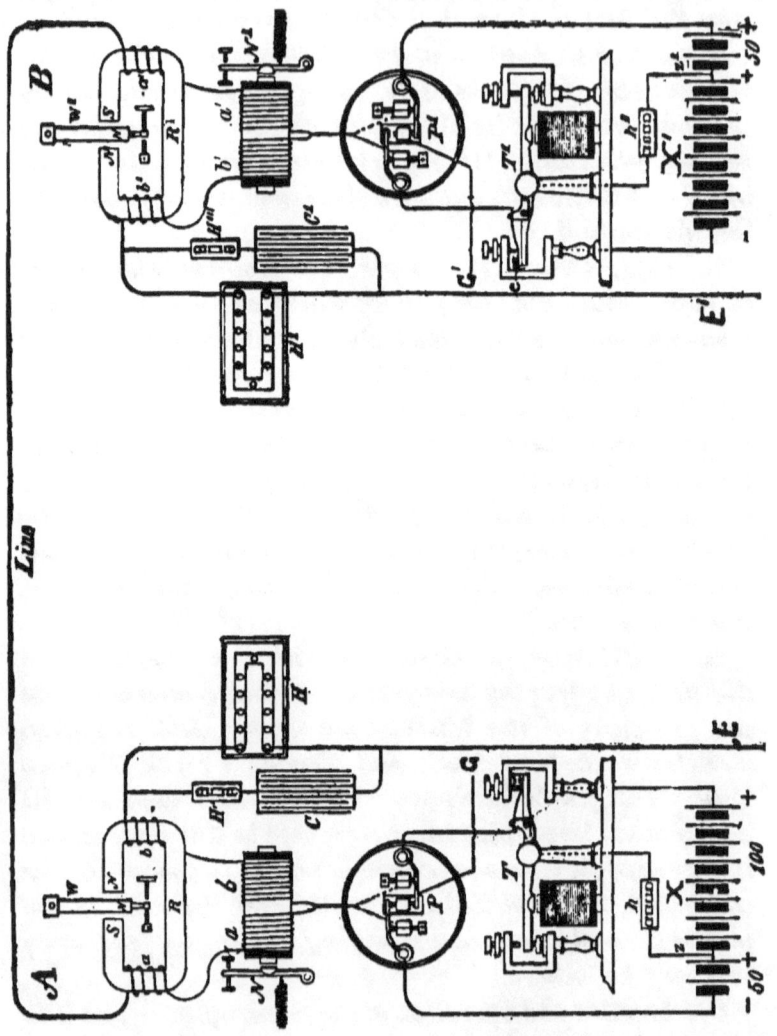

A and *B*. The Stearns, or *neutral*, relays, as they are called in distinction from the polar, are shown at N and N^1, and their connected transmitters at T and T^1; the

polar relays, at R and R^1, and their connected pole-changers at P and P^1. The artificial lines are constructed, as in the polar duplex, with small rheostats to regulate the action of the condensers. The main local batteries are shown at X and X', supposed, for convenience, to consist of 150 cells each, and to be tapped at z and z^1, so as to divide each into two batteries of 50 and 100 cells, so that the entire battery may be employed by a through connection, or only the smaller part, by a connection through the tap. The rheostats h and h^1, each having a resistance equal to that of the larger portion of the battery, are placed in the tap circuits, so as to maintain a constancy of resistance, equal to that of the entire battery, when only the smaller portion is in circuit.

The neutral relays are constructed with short cores, wound with coarse wire, and have strong retractile springs; the object being to reduce their sensitiveness: while the polar relays, whose permanent magnetism tends to make them sensitive, are constructed with long cores, wound with fine wire, to increase their sensitiveness; the resistance of the polar relay coil being about double that of the neutral relay coil.

This difference in sensitiveness tends to make the two relays independent of each other, so that the polar relay can be operated by a light current which does not affect the neutral relay; while the strong current required for the operation of the latter does not interfere with the simultaneous operation of the former.

The battery current passes through both the transmitter and the pole-changer, furnishing the constancy of current required by the polar relay, which is operated, as has been shown, by the change of polarity produced by the pole-changer, and, from its sensitiveness, responds with equal facility to the weaker current or the stronger; while the neutral relay, which requires intermissions in the current, responds only to the stronger current. And the battery connections of the transmitters, being arranged as shown, the weaker and stronger currents are alternately admitted to the line by the respective up and down movements of the transmitter levers, which alternately close the circuits through the smaller and larger batteries respectively.

The earth connections of the pole-changer levers are shown at G and G', and those of the artificial circuits at E and E'. The keys and sounders, with their local circuits, omitted in the diagram, are arranged in the usual manner; as are also the construction and winding of the relays, to which an ideal construction and winding are given in the diagram.

The keys at A being closed and those at B open, the current from the positive pole of battery X flows through the right hand post and tongue of transmitter T and lower tongue of pole-changer P, through both coils of neutral relay N and polar relay R with currents of equal strength; the right-hand current going through the artificial line to the earth at E, and the left hand current through the line to B; through coils a' of relays R' and N', the upper tongue of pole-changer P', the tongue and center post of transmitter T' to the negative pole z' of the smaller battery at X', through this battery to the positive pole, thence through the lower tongue and lever of P' to the earth at G'; from the earth at G, through the lever and upper tongue of pole-changer P, to the negative pole of battery X, and through the battery to the positive pole, completing the circuit.

There is now a 150 cell positive current, from battery X, traversing both coils of relays N and R equally, and hence producing no magnetic effect on them. But this current is increased in the left hand coils, a and a, by the 50 cell negative current of the small battery at X'; and this increase produces sufficient magnetic effect to attract the armature of relay R to the left hand stop, as shown, but not sufficient to attract that of N against the force of its spring. But this 200 cell current traverses the right hand coils, a' and a', of both relays at B, magnetizing their cores, so that the armature of relay N^1 is attracted to its left hand stop, as well as that of relay R^1 to its right hand stop, as shown, producing a down click in the connected sounder of each relay.

Now let the key of pole-changer P^1 be closed, reversing the polarity of the smaller battery at X', so that its positive pole is to the line, the current coming from the earth at G'. Its 50 cell positive current now flows through both coils of relays N^1 and R^1 equally, producing no magnetic effect; the current through the artificial branch going to the earth at E', while that through the line branch changes the inflowing 50 cell negative current to an outflowing 50 cell positive current; so that now the inflowing current through a' and a' of both relays is only that of 100 cells; but, on account of the opposite winding in each branch, this 100 cell inflowing current, on the right, and the 50 cell outflowing current, on the left, both magnetize each core in the same sense, so that both armatures are held in their former positions by the magnetism produced by a 150 cell current.

But the current through a, of relay R, being also thus reduced from that of 200 cells to that of 100, while the current through b still remains that of 150 cells, as before, armature W is attracted by the magnetism of a 50 cell current from the left to the right stop, producing a down click in the connected sounder, while the same preponderance of current does not affect the armature of relay N.

Now let the key of transmitter T^1 be closed, and the connection to Z^1 through the transmitter tongue, lever, and center post being opened at c, and that through its tongue and left hand post, to the negative pole of the full battery at X' closed; the 150 cell positive current of X' being now to the line, neutralizes the 100 cell positive line current from battery X, but adds a positive current of 100 cells to the 50 cell current through the artificial line at B; the armatures of both relays at B are therefore still held attracted as before. But the 100 cell line current through coils a and a at station A being neutralized, while the 150 cell current through b and b and the artificial line still remains, the magnetism in relay N becomes strong enough to attract the armature, producing a down click in the connected sounder, while the armature of relay R is still held on the right hand stop as before.

Now let the connections of battery X' be reversed by opening the key of pole-changer P^1, putting the negative pole to the line. The negative current of X', being now added to the positive current of X, produces a line current of 300 cells, which flows in through coils a' and a', holding the armatures of both relays at B attracted as before, being in the opposite direction to the former outgoing current through b' and b', which has ceased, and hence magnetizing the cores in the same sense. It also flows out through coils a and a, at station A, giving them a preponderance of 150 cells of current, over the current through b and b. The armature of relay N is therefore held attracted as before, while that of relay R is attracted from the right to the left stop, producing an up click in the connected sounder.

Now let the key of transmitter T^1 be opened, and the

result is a change of the negative connection of battery X' to that of the smaller battery at Z', restoring the conditions first considered, with both keys open at B and both closed at A; releasing the armature of relay N, which is pulled back by the spring, producing an up click in the connected sounder, all the other armatures remaining undisturbed.

All the conditions which can occur with both keys closed at one station, and one or both keys either opened or closed at the other station, have now been considered; the results being, evidently, practically the same with the operations at the respective stations reversed: and it has been shown that, in each case, both relays, at each station, can be operated without mutual interference; each responding to the manipulations of the key connected with the corresponding relay at the distant station, while the relays at the home station are unaffected by the manipulations of the home keys.

There are minor points of practical importance, in the adjustment, which require attention, to prevent false signals, liable to occur during the momentary change of battery connections, and which, even with the best regulation, cannot be wholly prevented, but are not of sufficient importance to prevent the practical operation of the system.

Repeating by the Quadruplex.—The quadruplex system is also employed for repeating in a similar manner to that of the repeaters already described; messages being repeated either from one relay to another of the same kind, or from one side of the system into the other, as from neutral relay to polar, or the reverse; the latter being the usual method.

Substitution of the Dynamo for the Battery.—The dynamo was first substituted for the battery in telegraphing by Stephen D. Field, and put in operation by the

Western Union Telegraph Company at New York in 1880, the Siemens-Halske dynamo being employed. But the result not being entirely satisfactory, the battery was reinstated in 1887. Meantime great improvement had been made in dynamo construction, and in 1890 the dynamo was again substituted for the battery, not only at New York, but at all the principal stations of the Western Union. The improved Edison dynamo was adopted; the 110 volt machines, grouped in series of five each, and operated at a working potential of 70 volts to each machine, being employed for the line circuits, and the 5 to 7 volt machines for the local circuits; a single dynamo being sufficient for the operation of several of the latter circuits, arranged in parallel.

Some changes are required to adapt telegraphic transmission to the dynamo current, especially in the use of the duplex and quaduplex systems. The reversals of polarity are made between two series of dynamos, one series furnishing the positive current and the other the negative. And the increase and decrease of current is produced by causing the currents to traverse routes of different resistance.

By the use of the dynamo a full supply of current can be obtained at far more economical rates than by the battery, and also greater constancy of current; the loss due to battery exhaustion being obviated.

The Wheatstone System of Automatic Rapid Transmission.—As telegraphic transmission by manual manipulation of the key does not exceed 25 to 50 words per minute, it becomes important in telegraph offices doing a large business to have some more rapid method. This is furnished by the system of automatic rapid transmission invented by Wheatstone, and employed in Great Britain for the Postal Telegraph, and in the United States by the Western Union at all its principal offices. Its relations to manual transmission are similar to those of printing to writing. It consists in the preparation of the message for transmission by recording it with perforations made in a strip of tough paper, and the subsequent automatic operation of a transmitter by this perforated strip, by which the message is transcribed in Morse characters on a similar strip by an inking register at the distant station.

Its instruments are a perforator and a transmitter, both of special construction, and an inking register of the ordinary type. The perforation of the paper is a purely mechanical operation. Three parallel rows of holes are perforated by punches which cut out the bits of paper; the central holes, which are smaller and less perfect than the others, being made in a continuous series, equally spaced, by the teeth of a star-wheel by which the strip is simultaneously drawn through, while the message is punched in the two side rows. The punches are operated by three keys pressed down by two little mallets in the hands of the operator against the force of spiral springs which bring them up again. The depression of the left-hand key punches a hole on each side of the central hole, preparing the paper for the transmission of a dot. The depression of the central key carries the paper forward one wheel-tooth space, making only the central hole, preparing the paper for the transmission of a space, the length of which can be doubled by two consecutive depressions of this key. The depression of the right-hand key carries the paper forward two wheel-tooth spaces, making a hole on the left when the first wheel-tooth space is passed, and on the right when the second is passed, each opposite a central hole, preparing the paper for the transmission of a dash. The appearance of a strip prepared in this manner for the transmission of the word "*operator*" is shown in Fig. 114, a care ul examination of which will verify the above description.

The paper, when thus prepared, is placed on the transmitter, and carried forward at a regular rate of speed by a star wheel whose teeth fit into the central holes, and which is operated by a spring or weight. Each side row of holes passes over the points of two vertical rods, connected with light apparatus contained in a box underneath, by which a pole-changer is operated. These points are pressed against the paper by spiral springs connected with the rods by bent levers, which operate the pole-changer; their upward movements being limited by two stops attached to an insu-

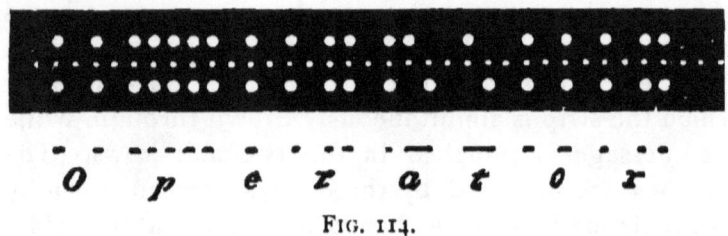

FIG. 114.

lating walking beam, to which a regular, alternate motion is imparted by the force which operates the star wheel. As each point meets a hole in the moving paper it passes up till its connected lever meets a stop on the walking beam below; and each alternate movement of this beam, made simultaneously with the advance of the paper one tooth space, pushes down the point which is up, and permits the other point to ascend till it is either stopped by the paper or passes up through a hole.

The ascent of the left-hand point through the perforated paper connects the positive pole of the battery with the line, bringing the pen of the register, at the distant station, into contact with the receiving paper;

THE ELECTRIC TELEGRAPH. 351

and the ascent of the right-hand point through the perforated paper reverses the polarity, withdrawing the pen from the receiving paper. And as the left point is adjusted one tooth space in advance of the right, if the two holes intended to transmit a dot pass over the points, the left point will ascend through its hole first, and be immediately depressed and the right point ascend through its hole; and hence the registering pen will touch the paper and be at once withdrawn, making the dot.

If now the paper move forward one tooth space, and the points meet no hole on either side, the negative pole being still to the line, and hence the registering pen still withdrawn, a space occurs on the receiving strip, the length of which depends on the distance the perforated strip moves before the left-hand point meets a hole. When this point meets a hole, the polarity being again reversed and a positive pole put to the line, the registering pen again touches the paper and is kept in contact till the right hand point meets a hole. If two tooth spaces are passed before this occurs, a dash is registered on the receiving strip, as shown at *a* or *t*, Fig. 114; but if only one is passed, a dot is registered, as before, as shown at *e*.

Hence it appears that the office of this perforated paper is simply to reverse the polarity, and that when the positive pole is put to the line, either a dot or a dash is registered according to the time elapsing before reversal; and when the negative pole is put to the line, either a short or a long space is registered according to the time elapsing before reversal. In this way dots, dashes, and long and short spaces can be registered automatically as rapidly as the instruments can be made to operate; the usual range of speed being from

125 to 250 words per minute, which is about five times that of manual transmission.

Hence one wire, by this method, can do the work of five wires by the ordinary method, and thus a large volume of telegraphic business be rapidly disposed of. This is especially advantageous in case of an accidental break in the connections, such as often occurs, since messages can still be received and prepared for transmission, and the accumulation rapidly despatched when the break is repaired. The simultaneous transmission of duplicate press despatches to numerous points from a central station can also, in this way, be greatly facilitated.

Submarine Telegraphs.—Submarine telegraph lines are constructed with cables like that shown in Fig. 115.

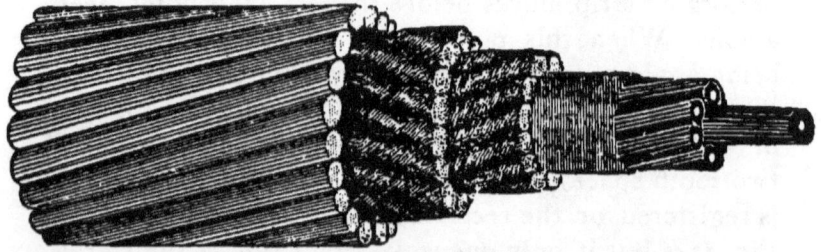

FIG. 115.

Seven or more No. 16 copper wires, thoroughly insulated, are inclosed in jute and protected by an exterior armor of iron wires wrapped with hemp; the interior being made water-proof. Deep sea cables, such as are used for the Atlantic lines, are made much lighter than those designed for shallow water, or shore ends, where the cables are more exposed to injury.

There are two important points of difference between the operation of long ocean lines, like those across the Atlantic, and ordinary land lines. One is, that the use of powerful electric currents, such as are required to operate the ordinary Morse instruments on long land lines, are liable to damage the insulating material of long submarine lines, and produce faults which soon render such lines worthless. The other is, that the peculiar conditions of great length, submergence in water, and an incasing armor of conducting material, insulated from the interior conductors, produces an excessive static charge, similar to that of the Leyden jar, which seriously impedes electric transmission.

Hence the first requisite is very sensitive apparatus, capable of responding to a low accompanying current, just sufficient to operate it, but not to injure the insulation ; such a current also reducing the static charge to its minimum. The second requisite is a condenser by which an induced, operating current is sent to the line and also a subsequent, reactive, opposing current, which neutralizes the return current of the static charge in the manner already described.

Thomson's reflecting galvanometer and accompanying scale, described in Chapter VI, page 130, furnishes the sensitive apparatus required, and is employed as the receiving instrument; the movements of the spot of light, on the scale, to the right or left of the zero mark, being made to indicate dots in one direction and dashes in the opposite direction, the pauses at zero indicating spaces.

Thomson's *siphon recorder*, also a very sensitive instrument, is employed for the same purpose. It consists of a capillary glass siphon, lightly poised, to which an oscillating, horizontal movement is given by apparatus operated by the line current through a local circuit, by which its point is made to vibrate, without contact, across a moving strip of paper, making a continuou., irregular line of dots with ink ejected in fine drops, drawn from a vessel in which the opposite end of the siphon is immersed; the ink and paper being oppositely electrified, and the various contortions of the line made to indicate the Morse characters.

Fig. 116 represents a cable with the transmitting

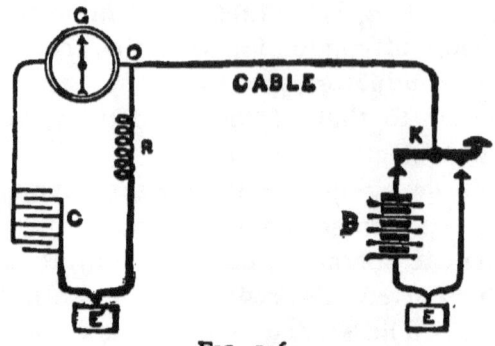

FIG. 116.

apparatus in connection at the terminal station on the right, and the receiving apparatus at the terminal station on the left, each kind being, of course, duplicated at each station, and the connections for transmitting or receiving changed by switches as required. The transmitting apparatus consists of the key K, permanently connected, by its rear contact, with the battery B, and arranged for connection by its front contact, when depressed, with the earth at E, with which the battery is also connected by its opposite pole. The receiving apparatus consists of the galvanometer G, connected with the earth at E', through the condenser C, on one side, and through the rheostat R, on the other. The condenser contains 40,000 or more square feet of tinfoil, and hence is capable of accumulating a very large charge, and the rheostat has a very high resistance.

The connection of the negative pole of the battery with the line through K, and of the positive pole with the earth, sends a positive current to the earth at E, and, as a result of the negative potential thereby produced in the line, a positive current flows from the earth at E' into the conductor C, repelling an equal amount of electricity from the opposite side of the condenser to the line and negative pole of the battery, completing the circuit. The line and line side of the condenser thus become negatively charged, but being at the same potential on both sides of the galvanometer G, this charge does not affect the needle after the first deflection produced by the charging; the needle therefore remains at zero. Now let the key at K be depressed, closing the front contact, and a positive current flows to the line, and into the line side of the condenser, from the earth at E, producing a deflection of the needle; but the condenser and line becoming charged positively, to the same potential, on both sides of the galvanometer, the needle returns immediately to zero. Now let the front contact at K be opened, and the former condition being restored, and the current reversed, a deflection in the opposite direction occurs, after which the needle returns to zero as before.

The advantage of a constant charge of the line in this manner becomes apparent when we consider that the complete charging or discharging of a submarine cable, two or three thousand miles in length, occupies several minutes, the time varying in proportion to the square of the length. The currents transmitted by the manipulations of the key rise and fall in waves; about $\frac{2}{10}$ of a second elapsing before any perceptible effect is produced at the opposite station by the closing or opening of a battery connection, and about 3 seconds being required for the wave to attain its full strength, and 3 more for it to decline. Hence, if the full phase of a wave intervened between signals, transmission would be exceedingly slow. But, by keeping the line charged, and employing the condenser, the current can be quickly reversed after each signal, without waiting for the wave to attain its full strength. And as the deflections indicating dots and dashes respectively are in opposite directions, their amplitude is of no consequence. Hence if, for instance, a succession of dots is required, as for the letter H, the first deflection has large amplitude, but is checked by a momentary reversal of current, and then further increased by another reversal; and thus, by four impulses in rapid succession, each less prominent than the preceding one, but all in the same direction, the four dots required are indicated. In like manner dashes are indicated by similar opposite deflections. A speed of 15 to 20 words per minute can thus be attained.

The constant charge of the cable and condenser in this manner reduces the interference of earth currents, often serious on lines of such length, to its minimum, so as to be of no practical importance ; the charge being sufficient to counteract the effect of such currents, under ordinary conditions, and leave a working surplus.

The earth connection at O allows a small percentage of current to pass through the rheostat, by which the potential on opposite sides of the galvanometer is equalized when the needle is at zero, without interference from the cable charge ; this potential being nearly the same as that of the line, on account of the high resistance of R.

Locating Faults.—When a fault or break occurs in a submarine line, it can be located approximately by the Wheatstone bridge test, invented by De Sauty. One arm of the bridge is connected with the cable, and the opposite arm with a condenser of known resistance ; and a current being sent through the instrument, flows into the condenser, through one side, and into the cable through the other, and to the earth at the fault ; and by this means the resistance from the shore to the fault can be ascertained. And the cable resistance per mile, in ohms, being known, the distance to the fault is easily ascertained.

As the resistance of the fault itself often varies considerably, it is important to test from each end and compare the results. And by subtracting the added results from the known resistance of the cable, the loss of resistance due to the fault can be accurately ascertained; and the proportionate amount of the remaining resistance from each end should give the resistance to the fault, from which the distance in miles can be ascertained as before.

When a cable contains separately insulated conductors, and a fault occurs in one of them, the loop method, which is considered one of the most accurate, can be adopted. Let the faulty wire be connected with a perfect wire at one end of the cable, and the two opposite terminals connected with the bridge at the other end; a current being transmitted, goes to the earth at the fault, passing round the loop from one side of the bridge, and direct to the fault from the other side. The known resistance of the cable being subtracted from the ascertained resistance of the loop, the remainder is the resistance to the fault from the end where the two wires are joined. The resistance from each end to the fault being thus ascertained, the distance can be calculated.

The Dial Telegraph.—In the dial telegraph, different forms of which have been invented by Wheatstone, Breguet, and Siemens, messages are indicated by ordinary letters displayed on a dial. In Breguet's receiving instrument the letters of the alphabet are arranged on the dial in a circle, around which a pointer rotates, stopping at the required letter. This pointer is operated by an electromagnet by means of clock-work in response to the closing or opening of the circuit by the transmitter; each make or break moving it one letter, always in the same direction; hence by a number of such changes in rapid succession it is brought to the required letter.

The transmitter has a similar dial around which a lever is moved, by which a toothed wheel is rotated, which either closes or opens the circuit alternately at each letter, by a connected lever, till the required letter is reached.

Such telegraphs are well adapted to the requirements of private lines, where the services of skilled operators are not available, and have been so employed in Europe; but, in the United States, printing telegraphs have been preferred for such lines.

Printing Telegraphs.—Telegraphs by which the message is printed in ordinary type have been invented by House, Hughes, Phelps, and Pope and Edison. A detailed description of these various instruments can be found in "Prescott's Electricity and the Telegraph" and similar books, to which the reader is referred. Their general principles of construction are similar to those of the dial telegraphs, and may be illustrated by supposing the lettered keys of a type-writer connected with telegraphic apparatus in such a manner that the depression of a key transmits a current by which a letter, corresponding to the one marked on the key, is printed on a strip of paper in the receiving instrument at the distant station, and the paper moved as required by the same apparatus. In fact, the printing telegraph is such a telegraphic type-writer.

CHAPTER XIII.

THE TELEPHONE.

Early History.—The reproduction of music and speech by electric transmission was first accomplished by Philipp Reis of Friedrichsdorf in 1861, though the transmission of sound and speech by the ordinary vibrations of a tightly stretched wire, as in the mechanical telephone, had been known for 200 years previous. The production of sound by electromagnetic vibrations was first observed by Page in 1837, in connection with the alternations of magnetism induced in an iron bar by an intermitting electric current in proximity; which, when occurring rapidly and rhythmically, gave rise to a musical tone. In 1854 Bourseul described a method which he had tried, by which speech could be electrically transmitted, which is practically the same as that now employed; and predicted its ultimate success when sufficiently developed. During the next twenty years the only progress made consisted in improvements of the musical telephone by various inventors.

About 1874 Elisha Gray began a series of experiments on the musical telephone, in Chicago, which led to the invention of a method by which it could be employed for the transmission of speech, for which, in 1876, he filed a caveat in the United States Patent Office. Meantime Alexander G. Bell had been making similar experiments, in Boston, expressly for the transmission of speech, and he also invented a method for which he applied for a patent on the same day that Gray filed his caveat.

A patent was granted to Bell March 7, 1876, his telephone was exhibited at the Centennial exposition at Philadelphia the same year, and its practical application to commercial use soon followed.

Claims for priority of invention were made by Gray, also by Daniel Drawbaugh of Pennsylvania, who claimed to have invented a similar telephone in 1872, and by Dr. Cushman, whose claims extended back to 1851. The litigation between Bell and Gray ended in a compromise, and that with the other parties was decided in Bell's favor. The transmission of speech by Reis's telephone was so imperfect as not to be considered entitled to priority as against the more perfect method invented by Bell.

Principles of the Telephone.—Sound, in the telegraph, is the arbitrary symbol of speech, while that in the telephone is its reproduction, the result of undulations of the air produced by the speaker at one end of the line, and reproduced by the transmitted currents at the other end. Tones, whether of music or articulate speech, are caused by the occurrence of such undulations in rhythmical order, their pitch depending on the number of undulations per unit of time, and their volume on the amplitude of those undulations. The property known as *timbre*, which distinguishes tones of the same pitch and volume from each other, depends on the manner in which the undulations are produced by different voices; a graphic representation showing undulations of different form.

In the telephone these three properties, pitch, volume, and timbre, are accurately reproduced by the undulations, so that the characteristic quality of the voice, and the manner of utterance, can be distinctly recognized, as well as the words spoken. And these undulations are produced by variations of current strength, and not by intermissions of current as in the telegraph.

It is the property of reproducing sonorous tones

which was first recognized, and caused the invention of the musical telephone to precede that of the speaking telephone, the properties of articulation and timbre being subsequently developed; and it is the development of these latter properties which distinguishes the Bell telephone from the musical telephones of Reis and Gray. As the Bell telephone embodies the leading principles of the musical telephones of Reis and Gray, a detailed description of the latter is unnecessary.

The Bell Telephone.—The Bell telephone is strictly a magneto-electric apparatus, generating its electric currents by the movements of a magnetized armature in proximity to a conductor forming a closed circuit. The construction of its principal instrument, employed now as a receiver only, but formerly both as a transmitter and a receiver, is shown in Fig. 117. A round, hard-rubber case, $6\frac{1}{2}$ inches long and $1\frac{1}{2}$ inches in diameter, enlarged at one end as shown, incloses a round bar magnet, $\frac{1}{4}$ of an inch in diameter, to whose north pole, N, is attached a wooden bobbin, bb, wound with fine, silk-covered, copper wire, whose terminals, gg, are attached to larger wires, cc, connected with the binding-posts hh. A very thin sheet-iron disk, PP, $2\frac{1}{4}$ inches in diameter, varnished to protect it from oxidation, covers the circular space within which the bobbin and magnet pole are placed, and is held in position, at its edges, by the ear piece VV, which is screwed over it as shown. The center of this disk comes close to the magnet, the distance being adjusted by the screw d; sufficient space being allowed for a slight vibration of the disk, without con-

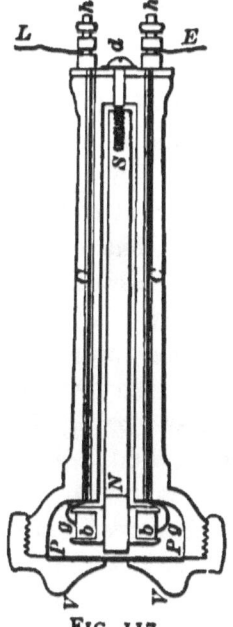

FIG. 117.

tact. There is also a similar amount of space between the disk and the ear piece: and the vibrations produced by the variations of magnetic energy are limited by the elasticity of the disk; its center vibrating very slightly, while its edges are held fast. This disk is known as the *diaphragm*.

In the bottom of the cavity of the ear piece, opposite the center of the diaphragm, is an opening, $\frac{1}{18}$ of an inch in diameter, through which the undulations produced in the air by the vibrations are transmitted to the ear.

If one terminal of the coil, as L, be connected with an ordinary telegraph line, and the other, E, with the earth, and corresponding connections be made with a similar instrument at the opposite end of the line, one instrument can be used as a transmitter and the other as a receiver. When a person speaks into the transmitter, the undulations of the air cause the diaphragm to vibrate ; each vibration varying the distance between the disk and the magnet, producing corresponding variations of magnetism in the disk as an armature, which induce an alternating current of varying strength in the coil. This current, transmitted to the coil of the receiver, at the opposite end of the line, reproduces like variations of magnetism in its diaphragm, and hence corresponding vibrations and undulations, which reproduce the words spoken into the transmitter.

The improved telephone, as above described, was patented by Bell Jan. 30, 1877; the chief claims of the patent being the substitution of the iron disk for the stretched animal membrane previously employed, and the magneto-electric current for the battery current.

Improved Transmitters.—It is evident that much of the energy of the transmitted voice is spent in overcoming the various interposed resistances, so that when heard in the receiver it is comparatively weak, and the feebler tones are liable to be indistinct. Hence various means have been devised to counteract the effects of this consumption of energy and render the delivery more distinct.

The Edison Transmitter.—It was observed by Du Moncell that increase of pressure reduces the contact resistance between conductors ; and that this effect is increased by reduction of hardness and increase of electric resistance in the conductors themselves: and hence that variations of current strength may be produced in this way. It was also observed by Edison that carbon is peculiarly adapted to fulfill these conditions; and in accordance with these observed facts he constructed, in 1878, the first transmitter in which carbon was employed. He also employed platinum, as proposed by Gray, using a disk of each material in contact, and producing a slight contact between the platinum disk and the vibrating diaphragm by an ivory button; the high resistance of each substance, the difference of their hardness, and the varying pressure produced at the numerous contacts by the vibrations of the diaphragm, being intended to improve the conditions of transmission.

Edison also employed an induction coil, as had previously been done by Gray to increase the E. M. F. of the line current; connecting its primary coil with the circuit of a local battery, whose current traversed the transmitter, and its secondary with the line, by one terminal, and with the earth by the other, completing the circuit by corresponding connections of a Bell receiver with the line and earth at the opposite station. By this means the relative conditions of E. M. F., resistance and current strength, in the two circuits, could be varied, so that a large current of low E. M. F. and resistance, in the primary, could be converted into a small current of high E. M. F., in the secondary, capable of overcoming the line resistance, and producing sufficient amplitude of vibration in the diaphragm of the receiver to render the tones audible and distinct.

The Blake Transmitter.—The transmitter invented by Blake, and now employed by the Bell Telephone Company throughout the United States, is an improved form of the Edison transmitter. A vertical section of its principal parts is shown in Fig. 118, attached to the door of a little cabinet in which it is contained; Fig. 119 giving a rear view, with the door open, showing also the induction coil and connections.

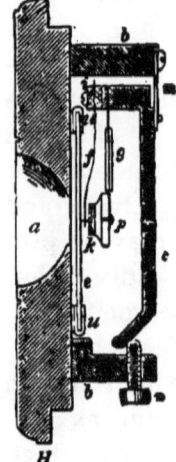

Fig. 118.

Opposite the mouth-piece *a*, formed

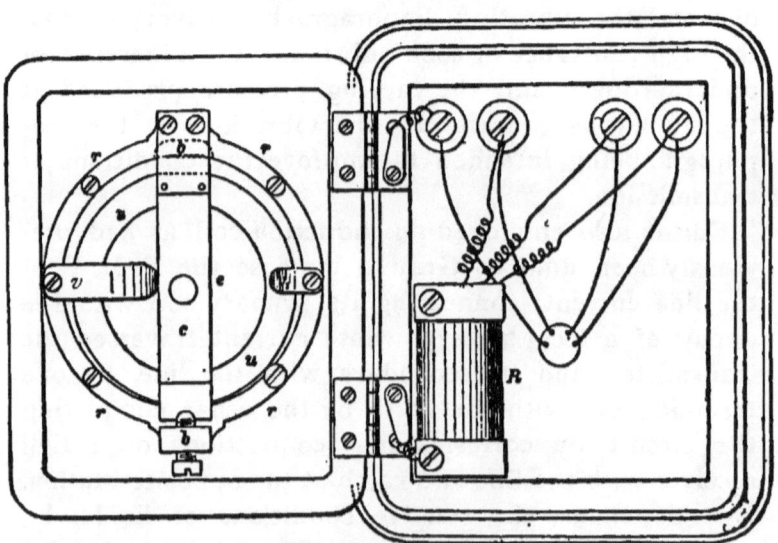

Fig. 119.

by a funnel-shaped opening in the door, is fixed the sheet-iron diaphragm *e*, supported by a soft-rubber ring, *uu*, and the springs v v_1; v_1 pressing on the ring, and v on the diaphragm itself, a piece of felt attached to the spring v intervening at the point of contact. This ring insulates the diaphragm from the conductors both of sound and electricity with which it is surrounded; and the felt on the spring prevents excessive vibration, reducing it to the normal quantity required for distinct transmission. An iron ring, *rrrr*, has two projections, *b b*, each about ¾ of an inch in length, from the upper of which the iron bar, *c*, is suspended by the brass spring *m*. A flat, metal spring *g*, attached to the short upper arm of *c*, and inclosed in soft rubber to lessen its vibration, carries, at its lower end, a brass disk *p*, to which is attached a carbon disk *k*, both suspended opposite the center of the diaphragm *e*. A thin, flexible steel spring *f*, insulated from *g* by the block of vulcanized fiber *i*, is also attached to the upper arm of *c*, its lower end having two platinum points, one of which makes contact with the center of the plate *k*, and the other with that of the diaphragm *e*. The tension of the springs *f* and *g*, and hence the pressure of the various contacts between the disks, is adjusted by the screw *n*, by means of its bearing against the lower, bent arm of *c*.

Accessory Apparatus.—A Leclanché cell, in the lower compartment of the cabinet, supplies the current, the connections being through the primary circuit of the induction coil *R* to one of the door hinges, thence to the ring *rrrr*, upper projection *b*, upper arm of *c*, springs *m* and *g*, plates *p* and *k*, platinum point, spring *f*, through a lever in the upper compartment of the cabinet, and thence through the other door hinge to the opposite pole of the battery. The secondary circuit of the coil *R* is connected with the line and the earth by its opposite terminals.

The lever referred to is used to open and close the

primary circuit. It terminates in an exterior forked hook in which the receiver is hung. The weight of the receiver depresses the lever in opposition to the force of a spring and opens the primary circuit, thus preventing exhaustion of the battery by the generation of current when not required. The receiver is connected with the line by a flexible conducting cord, and when taken off the hook to be applied to the ear, the lever, relieved of its weight, springs up and closes the battery circuit, sending a current through the transmitter and primary circuit of the induction coil, by which an induced current is generated in the secondary circuit and connected line. The message is then spoken before the mouthpiece of the transmitter.

The Signaling Apparatus.—The signaling apparatus, invented by Gilliland, is also contained in the upper compartment of the cabinet and connected directly with the line. Its transmitter is a magneto-electric machine, operated by an exterior handle, by which a line current is generated which drops an annunciator at the central station. Its receiver is a *call-bell*, constructed with two gongs, between which the clapper vibrates. The latter is operated by a double coil electromagnet, to the armature of which it is attached. This armature is mounted a short distance in front of the poles, being pivoted at its center on one of the poles of a bar magnet, mounted between the coils. It has therefore the same polarity as the pole with which it is connected; hence when an alternating current, transmitted from the central station, passes through the coils one end of the armature is attracted and the other end repelled by each pole alternately, as the polarity changes, producing a rapid, rocking motion which brings the clapper alternately into contact with each gong.

The Exchange.—A telephone exchange is a central station in which intercommunication between all the subscribers' stations connected with the net-work of lines belonging to any particular local system is established. Each line radiates from this exchange, and all communication between subscribers must pass through it. Where the system is very extensive, embracing several thousand subscribers, there are also several district stations between which there are trunk lines by which connection between subscribers in different districts can be made. The terminals of all the lines are connected with a switch-board, which thus often becomes a very extensive and elaborate apparatus.

The Multiple Switch-Board.—A section of a multiple switch-board, such as is now employed by the Bell Telephone Company in the United States, is shown in Fig. 120. The supporting plates are composed of insulating material and pierced with holes through which connection is made with the various lines and connected apparatus attached to its back. Lady operators sit in front of it, each having a receiver to her ear and a transmitter suspended before her by which she can communicate with all the subscribers in her section. It is divided into compartments, *A*, *B*, *C*, and *D*.

Compartment *A* contains the annunciators which indicate the calls from subscribers, to each of whom one is assigned. They are constructed with hinged brass tablets, each having the subscriber's number on its inner face, and connected with an electromagnet by which a spring catch is operated, which keeps the tablet closed when not in use. When a subscriber transmits a current from his electromagnetic machine, it passes through the coil of this magnet, the armature is attracted, lifting the catch and releasing the tablet, which

368 DYNAMIC ELECTRICITY AND MAGNETISM.

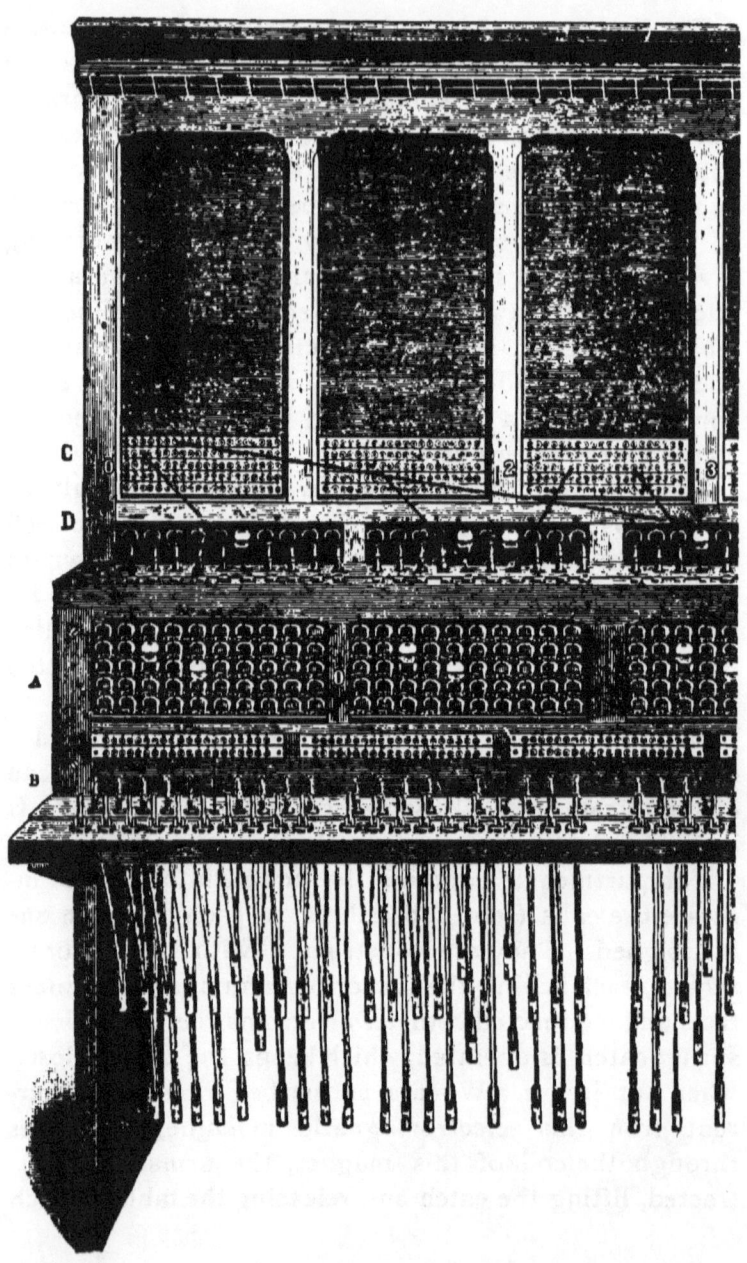

FIG. 120.

drops into a horizontal position and displays the subscriber's number.

In compartment B are shown the flexible conducting cords for making connection with the subscribers' lines, kept stretched by suspended pulleys to prevent entanglement; each attached to a plug shown in the rear row, on the shelf, and passing through a connection controlled by a cam shown in the front row. Each of the holes, shown above the row of plugs, connects with a spring-jack in the circuit of each subscriber included in that section; while each of the holes in compartment C connects with a spring-jack in the circuit of each subscriber in the entire system; each wire in compartment C passing through a spring-jack in every section. Hence each operator can connect, in her own section, any of her subscribers with any subscriber belonging to her own or any other section, through compartment C; while in compartment B, she can connect herself only with the subscribers included in her own section.

In compartment D are the "clearing out" annunciators by which either of two subscribers, in communication, can indicate the close of the conversation by a signal current which drops the connected annunciator in this compartment. On the shelf in this compartment are shown a row of plugs, each attached to the same cord as the corresponding plug in compartment B, so that any subscriber connected with a cord by the insertion of one of the lower plugs in the hole bearing his number in compartment B, can be connected with any other subscriber by insertion of the upper plug connected with that cord in the hole bearing the other subscriber's number in compartment C, as stated above. In front of these plugs is shown a row of buttons, each connected with a contact by which a circuit can be closed between any subscriber with whom connection is made in compartment *C* and a magneto-electric machine, and a call rung on the subscriber's bell; a number of these machines being kept in constant operation, at a large station, by a water motor or otherwise.

When an annunciator drops, the operator places a plug, taken from the row in compartment *B*, in a hole in one of the vertical rows immediately above it having the corresponding number, at the same time turning down the corresponding cam lever, so as to close the subscriber's circuit, and inquires what number is wanted; having ascertained this, she applies the terminal plug of her receiver to a special tube connected with the line of the subscriber called for, and if the contact produces noise in her receiver, it indicates that there is a current on the line, and she informs the subscriber that the person wanted is conversing with some one else. But as soon as she ascertains that the conversation has ceased, as indicated by silence when contact is made, she takes a plug from the row in compartment *D*, connected with the same cord as the plug inserted in the hole bearing the calling subscriber's number in compartment *B*, and inserts it in the hole, in compartment *C*, bearing the number of the subscriber called for, which puts the two in communication; and having rung a call to the subscriber wanted, by pressure on the signal button, she opens the cam connection and closes the annunciator.

All the spring-jacks through which any subscriber's line passes are in connection through a separate wire, and it is on a connection with this wire that the test above described is made.

As each clearing-out annunciator is in circuit only when two subscribers are connected and can be operated by either, a comparatively small number is sufficient; but each subscriber being constantly in circuit with one of the other annunciators, there must be as many of these as there are subscribers.

Hughes' Microphone.—In 1878 Hughes invented the instrument known as the microphone, by which feeble sounds can be transmitted, and reproduced, greatly amplified, in a telephone receiver. In his first experiments a wire nail, making loose contacts with two other nails, was employed, but when the superior quality of carbon for telephonic transmission was demonstrated by Edison, Hughes substituted a small carbon rod, pointed at both ends, and loosely mounted vertically between two carbon supports attached to a thin sounding board. The terminals of a battery circuit, in which the receiver is included, are attached to these supports, and the slightest sound made near this simple instrument, as the ticking of a watch or the walking of a fly, can be distinctly heard in the receiver, at a distance of several feet.

The instrument is too sensitive to be used for ordinary transmission, but the advantage of loose contacts, as thus demonstrated, has been utilized to increase the sensitiveness of carbon transmitters, by the substitution of granulated carbon for carbon plates.

Theory of Telephonic Transmission.—Opinion is divided in regard to the fundamental principles of telephonic transmission, and especially in regard to the functions of carbon as a transmitter. It is well known that heating reduces the electric resistance of carbon, and hence it is maintained that the variation of heat generation in the carbon, produced by the variation of pressure due to the loose contacts, produces a corresponding variation of resistance and hence of current strength. While the heat thus generated must be almost infinitesimal in quantity, nevertheless its ratio to the molar and molecular vibrations, in an apparatus of such delicate sensitiveness as the telephone, is believed to be sufficient to account for the improved transmission; and observation shows that continuous use produces a perceptible rise of temperature in a transmitter. This theory applies more particularly to the microphone, but its application to carbon transmitters in general is obvious.

The generally accepted theory of molar vibrations of the diaphragms, as already explained, is disputed by some, who maintain that telephonic transmission is chiefly, if not wholly, due to molecular vibrations produced by the variations of magnetism. In proof of this it is shown that such transmission is possible with instruments constructed with thick disks, incapable of the molar vibrations ascribed to the thin ones. This theory also receives confirmation from the slight increase of length shown to be produced in a steel bar by magnetizing it, indicating that variations of magnetic strength in a magnetized bar must produce corresponding variations of length. The click due to magnetization, as observed by Page, also shows that magnetic, molecular vibrations may become sonorous.

It is probable that all these theories are more or less applicable; that molar vibrations, molecular vibrations, and variations of resistance due to variations of temperature, all contribute to the observed results. The circular shape of the diaphragms and plates is also important in contributing to evenness and regularity of vibration, and hence producing corresponding evenness and regularity in the atmospheric undulations, and should not be overlooked.

Multiplex Telephony.—The ordinary conditions of telephonic transmission require that each subscriber should have a separate wire, since each must operate his own line, on which strict privacy is required, and be in constant connection with a central exchange through which he can be put in communication with others; while, in telegraphic transmission, the messages of numerous persons are sent, in rotation, from a central station, over the same wire, by experts, to whom alone their contents are known. Hence a telegraph line can be kept constantly occupied by thousands of persons, and rapid, automatic transmission employed, while a telephone line is occupied, usually, only a small proportion of the time by a single person.

As this difference makes the expense of telephonic transmission enormous, as compared with telegraphic, various methods have been devised for simultaneous duplex telephonic transmission, similar to that of telegraphic, and also for the occupancy of the same line by several persons in rotation. While experiments of this kind have been, to some extent, successful, their success has not been such as to warrant their general, practical adoption.

The occupancy of the same line by different subscribers in rotation is, however, in practical use for intercommunication between different central stations; the trunk lines, already referred to, being employed in this way; the proportionate number of such lines to the subscribers' lines being dependent on the amount of intercommunication.

A description of the various experimental methods, referred to above, may be found in Preece and Maier's book on " The Telephone."

Long Distance Telephony.—The multiplicity of lines required for the operation of a practical telephone system, as shown above, and the difficulty of overcoming resistance and induction so as to reproduce speech distinctly at the terminals of long lines, has till recently confined the use of the telephone chiefly to the limited areas of towns and cities. Experiments in long distance telephony were formerly made on telegraph lines; and as these were composed of single, grounded, iron wires, arranged in numerous parallel lines, in close proximity on the same poles, and hence subject to high mutual induction, the results obtained were not sufficiently encouraging to induce the investment of capital in independent telephone lines, and the true causes of failure were not at first clearly perceived.

The effects of induction beween parallel telephone lines in proximity, which causes a person with a receiver to his ear, waiting to be put in communication, to hear, indistinctly, conversation between others, is well known. But the fundamental difference between telephonic and telegraphic transmission, as already shown, aggravates this inductive effect, when lines used for both purposes are in proximity; the strong, intermittent current on the telegraph line overcoming the light, undulatory current on the telephone line to such an extent as to interrupt the undulations and render transmitted speech indistinct, especially on long lines; producing a continuous, accompanying crackle.

Van Rysselberghe's System.—This effect can be reduced by giving the telegraphic current an undulatory motion similar to that of the telephonic. This has been done by Van Rysselberghe, a Belgian electrician. He introduced two electromagnetic primary coils, having iron cores, into the telegraph circuit, one between the battery and the key and the other between the key and the line, and passed the ground wire through a condenser. The rise and fall of the current at each intermission, caused by the charge and discharge of the condenser, in connection with the retardation due to self-induction and magnetic lag in the coils, produces an undulatory effect, similar to that in the ocean cable lines. By connecting the telephone apparatus with the line through a separate condenser, the same line can be used for simultaneous telegraphic and telephonic transmission.

As this system requires that all parallel lines, mounted on the same poles, shall be constructed in this manner, and shall also have special apparatus to prevent the telephonic induction referred to above, all of which entails considerable extra expense ; and as it also retards telegraphic transmission, it has not come into general practical use, though employed on some of the Belgian lines.

The American System.—In the long distance telephone system now employed by the American Telephone and Telegraph Company the lines are composed of No. 12 copper wire, and are complete metallic circuits, without ground connections ; each line having two wires, on each of which the current flows in opposite directions. The superior conductivity of copper, as compared with iron in one branch of the circuit, and the earth in the other, is apparent ; besides which the mutual induction of opposite currents in the two parallel lines proportionately increases the current strength in each direction ; while the freedom from grounded connections at the terminals prevents interruption from the inductive effects of contiguous, grounded, telegraph and telephone lines, usually numerous at such terminals.

Where several such lines are mounted on the same poles, the mutual induction between currents, which would produce " cross talk " between the lines, is neutralized by introducing lines constructed with numerous transpositions between the straight lines. These transpositions are made at regular intervals of a few poles apart, by crossing the two branches of the line, without contact, so that each takes the place of the other on the cross-arms. By this means the adjacent lines, on either side, are alternately brought into proximity, through short sectional distances, with wires bearing reversed currents in each section, and thus the effects of induction are neutralized.

The instruments employed are the Hunning transmitter and the Bell receiver, with the signaling apparatus already described.

The Hunning Transmitter.—The diaphragm of this transmitter is a disk of platinum foil, supported by a metal ring, and protected by wire guards in front. A thicker disk of brass, gold-plated, is placed back of this one, and parallel with it, at a distance of about $\frac{3}{32}$ of an inch, and the space between filled with finely granulated carbon, sifted free from dust, whose superior quality as a transmitter has been already referred to. The whole is inclosed in a wooden box, to which is attached a metallic, funnel-shaped mouth-piece, in front, and, at the back, are attached the binding-posts for the battery terminals, one connected with the supporting ring of the diaphragm, and the other with the rear plate; so that the current must pass through the carbon.

Transmission on Long Distance Lines.—Lines 600 miles long are now in practical working order, speech being reproduced with distinctness, and lines 1000 miles in length are projected. On the line between Chicago and Milwaukee, 90 miles in length, whispered conversation and the ticking of a clock can be reproduced distinctly, also a hiss, the most difficult sound to transmit by the telephone.

For the accommodation of the Bell Telephone Company's subscribers, connections are provided at the central stations between their lines and the long distance lines, and the extra price for such service charged to the subscriber's account whenever such connection is made. But the reproduction of speech through such connections, is not so perfect as by direct connection.

INDEX.

A.

	PAGE
Absolute Magnetic Intensity	43
Accessory Apparatus, for the telephone	365
Accumulator	235
Action, heat developed by electrochemical	254
", local, in the battery cell	9
Advantages of the Alternating Current Dynamo	189
Agonic Line	38
" Map of the United States	52
Agitation of the Solution, in electroplating	226
Alliance Machine, the	166
Alternating Current Dynamo, advantages of the	189
" " ", the Gordon	185
" " ", " Westinghouse	186
" " Dynamos	185
" " ", separate excitation	189
" " Motor, the	198
Aluminium, Bunsen's process for	229
", St. Clair Deville's process for	229
", the Hall process for	231
Amalgamation of the Zinc, in battery cells	8
American Morse Code	312
" System of long distance telephony	375
Ammeter, the Weston	135
Ammeters, gravity	145
", voltmeters and	134
Ampere, the	6, 116
Ampere-Hour, the	117

INDEX.

	PAGE
Ampere's Table	82
" Theory of Magnetism	89
" Rule	73
Analogy between Magnetic and Electric Phenomena	67
Analyzer, the, in the polarization of light	284
Angles, measurement of	122
Angular " " Deflective Force	123
Anions	207
Annual and Diurnal Variation	50
Anode, defined	207
Anodes, the, in electroplating	221
Apparatus, accessory, for the telephone	365
" , signaling " " "	366
Arc, the	297
" Lamp, the	297
" Light, "	295
" , multiple, defined	29
Armature, of the magnet	55
" , " " electromagnet	78
" , " " dynamo	170
" , " " " , the cylinder	174
" , " " " , Gramme, interior wire of the	173
" , " " " , the Pacinnotti-Gramme	170
" , " " " , " Siemens	167
Armatures, of the dynamo, closed circuit and open circuit	176
Armature's Magnetic Poles, location of the, in the dynamo	177
Arrangement, station, in the telegraph	321
Artificial Magnets	54
Attraction and Repulsion, polar	58
Astatic Galvanometer	129
" Needle, the	73
Automatic cut-out, for arc lamps	303
" Rapid Transmission in telegraphy, the Wheatstone system	348
" Regulation, in the arc lamp	300
Auxiliary Operations, in electroplating	223
Ayrton and Perry's Spring Voltmeters and Ammeters	142

B.

Balance, Coulomb's torsion	67
Ballistic Galvanometer	134

INDEX. 381

	PAGE
Battery, the, for the telegraph	314
" , cell, element and	3
" , De La Rive's floating	84
" Formation	28
" " , Two-fluid cells	23-34
" , Grove's gas	233
" , Sign	3
" , substitution of the dynamo for the, in telegraphing	347
" , the voltaic. Definitions	1-12
Becquerel's Discoveries in the electric relations of light	288
Bell Telephone, the	361
Bichromate Cell, potassium	15
Bifilar Suspension	129
Biot's Law	44
Bodies, paramagnetic and diamagnetic	64
Blake Transmitter, the, for the telephone	364
Blasting, electric	257
Break, and make, in the electric circuit	93
Bridge, the Wheatstone	157
Brushes, dynamo	170
" " , position of the	178
Buckling, conductivity and, in storage cells	246
Bunsen Cell, the	26
Button Repeater, the, in the telegraph	325

C.

Calibration of Galvanometer	124
Callaud Cell, the	25
Candles, electric	295
Capacity of conductors, electro-thermal	255
" , storage, of storage cells	249
Carbons, arc lamp	300
Cardew Voltmeter, the	147
Cations	207
Cathode "	207
Cause of Deflection, of the needle	74
Cautery, electric	258
Cell, the Bunsen	26
" , " Callaud	25
" , " Clark	139, 142
" , " Daniell	23

INDEX.

	PAGE
Cell, diamond-carbon	20
" , Element, and Battery	3
" , the Faure Storage	236
" , " " " , defects of	240
" , " " " , improved	241
" , " " " , electric energy of	244
" , " Grenet	16
" , " Grove	26
" , " Julien storage	247
" , " Law	20
" , " Leclanché	17
" , " mercuric bisulphate	17
" , Planté's secondary	236
" , " " , electric energy of	239
" , potassium bichromate	15
" , the Pumpelly storage	247
" , " silver chloride	27
" , Smee's	13
" , theory of electric generation in the	6
" , the voltaic, operation of	6
" , Walker's	14
Cells, connection between	33
" , dry	21
" , durability of storage	249
" , gravity	25
" , one-fluid	13–22
" , polarization of one-fluid	22
" , two-fluid. Battery formation	23–34
" , " " , construction of	23
" , weight of storage	247
" , zinc-carbon	13
Charge and Discharge, effects of, on storage cell plates	245
Charging and Discharging storage cells, relative time of	249
Chemical Equivalence	216
" Reaction in the Faure cell	240
" " " " Planté "	237
Circuit, open, defined	19
Clamping, insulation and, in the battery	9
Closed Circuit and Open Circuit Armatures	176
Code, the American Morse	312
" , " International Morse	312

INDEX. 383

	PAGE
Coefficient of Magnetic Induction	76
" " Magnetism	98
" " Mutual Induction	95
Coil, induction	98
" a Converter, the	105
" , spark	109
Coils, resistance	155
Common Galvanometers	134
Commutation	165
Commutator, dynamo	170
" , " , improved	171
" , Ruhmkorff's	104
Compass, the mariner's	35
" , " surveyor's	37
Compensating Magnet	74
Composition of Grids, for storage cell plates	247
Compound Winding, in the dynamo	183
Compounds, electrolysis of mixed	212
Condenser	99
" , Leyden Jar and	233
" , " " as a	102
" , operation of	102
Conditions of Electric Energy in the battery cell	4
" " Electrolysis	210
Conductivity and Buckling, in storage cells	246
" , insulation and	112
Conductor	113
Conductors, electro-thermal capacity of	255
Connection Between Cells	33
Connections, repeater, telegraph	327
Consequent Poles in the magnet	61
Constant Current Dynamo	183
" Potential "	183
Construction of Core in induction Coil	101
" " " " dynamo armature	174
" " " " " field-magnets	179
" " " " electromagnets	76
" , line, in the telegraph	320
" and Operation of the Quadruplex	341
" special, in induction coils	102
" of Two-Fluid Cells	23
Convection, effect of in electrolysis	218

384 INDEX.

	PAGE
Converter, the	189
" , the coil a	105
" , " Tesla Motor as a	202
Core of dynamo armature	174
" " " field-magnets	179
" " electromagnet	76
" " induction coil	97, 98
" " " " , construction of	101
" " " " , induction of	97
" " " " , sliding	101
Cosine	122
Cosmic Variation	51
Coulomb, the	117
Coulomb-Meter, the Forbes	150
Coulomb's Torsion Balance	67
Couronne de Tasses, the	2
Crater and Point, the, in arc-light carbons	298
Current, deflection by the electric	71
" , direction of, in the dynamo	173
" , electric	5, 114
" , " , deflection of by the magnet	82
" and Electrylote, relative conditions of	219
" , establishment of, in the arc lamp	299
" , extra	96
" , faradic, physiological effects of	107
" , generation of, dependent on variation of intercepted magnetic force	95
" Induced by Another Current	91
" " " Opening or Closing Primary Circuit	93
" " " Magnet	90
" " " Varying the Strength of Primary Current	94
" Induction, results of	94
" " , rotary movement by	87
" Meter, the Edison	150
" , position and, in the incandescent lamp	305
" Reversal, effect of, in electrolysis	217
Currents, eddy	56, 196
" , electric, generation of by induction	90
" , Foucault	56, 196
" , mutual induction of electric	84
Cut-out, automatic, for arc lamps	303

	PAGE
Cut-out, Ground-Switch, and Lightning Arrester, for telegraph...	319
Cylinder, armature, in the dynamo..........................	174

D.

Daniell Cell, the..	23
Declination...	37
Defects of the Faure Cell,...................................	240
Definitions. The voltaic battery.............................	1–12
Deflective Force, angular measurement of.....................	123
Deflection, cause of...	74
" , by the Electric Current............................	71
" , of " " " by the Magnet..............	82
" , magnetic force ascertained by......................	42
De La Rive's Floating Battery................................	84
Deposit, thickness of, in electroplating......................	225
De Tasses, the Couronne.....................................	2
Details of electroplating, various............................	220
Development of the Electric Motor...........................	190
Diagrams, thermo-electric....................................	266
Dial Telegraph, the..	357
Diamagnetism, experiments in	79
Diamagnetic Bodies, paramagnetic and........................	64
" and paramagnetic substances, list of..............	81
Diamond-Carbon Cell, the...................................	20
Differential Galvanometer...	133
" Relay...................................	333, 336
Different Kinds of Electric Measurement.....................	118
Dip, inclination or...	40
Diplex Transmission, in telegraphy	340
Dipping Needle, the..	40
Direction of the Current, in the dynamo......................	173
Discharge in Air and in Vacuo................................	107
" , charge and, effects on storage cell plates............	245
" , E. M. F. of, in storage cells.......................	246
Discharging Storage Cells, charging and, relative time of.......	249
Discoveries, Becquerel's, in the electric relations of light......	288
" , Faraday's " " " " " "	284
" of Galvani " " battery......................	1
" , Kerr's, " " electric relations of light........	289
" , Kündt & Röntgen's, " " " "	289
" , Verdet's, in the " " " "	287

INDEX.

	PAGE
Discoveries of Volta, in the battery	2
Disque Leclanché Cell	19
Distribution, elevated road electric	204
" , magnetic, lamellar	63
" , multiple series and series multiple	307
" of electric Power	203
" , parallel, in electric lighting	305
" , series, " " "	303
" , three wire system of, in electric lighting	308
Double Reflection, effects of, on magneto-polarized light	292
Dry Cells	21
" Pile, Zamboni's	7
Duplex Telegraphy	329
" , the polar	333
" , " " , operation of	338
" , " Stearn's	330
Durability of Storage Cells	249
Dynamic Electricity Defined	1
Dynamo, the	168
" , advantages of the alternating current	189
" Brushes	170
" Commutator	170
" , constant current	183
" , " potential	183
" , the Edison	185
" , " Gordon	185
" and Motor, the	165–205
" , the, as a Motor	193
" , " , substitution of, for the battery in telegraphing	347
" , " Westinghouse	186
Dynamos, alternating current	185

E.

	PAGE
Early History of the telegraph	310
" " " " telephone	359
Earth's Magnetic Poles, the	37
Eddy Currents	56, 196
Edison Current-Meter, the	150
" Dynamo, the	185
" Transmitter, the, for the telephone	363
Effect of Convection in electrolysis	218

INDEX. 387

	PAGE
Effect of Current Reversal in electrolysis	217
" , the Peltier, in the electric relations of heat	268
" , " Thomson, " " " " "	269
Effects of Charge and Discharge on storage cell plates	245
" " Double Reflection on magneto-polarized light	292
" , Magnetic, in electrolysis	216
Electric Blasting	257
" Candles	295
" Cautery	258
" Current	5, 114
" " , deflection by the	71
" " , deflection of, by the magnet	82
" Currents, generation of, by induction	90
" " , mutual induction of	84
" Energy, conditions of, in the battery cell	4
" " of improved Faure cell	244
" " , loss of, in the dynamo	195
" " required for electroplating	225
" Fuses	258
" Gas-Lighting	109
" Generation in the Cell, theory of	6
" Heat to Electric Light, the relations of	279
" Horse-Power, the	118
" Intensity	29, 32
" Lighting	295
" Measurement	110–164
" " , different kinds of	118
" Motor, development of the	190
" Perforation	107
" Potential	110
" Preparation of Plates for Faure Cell	244
" " " " " Planté "	236
" Pressure	111
" Reduction of Ores	228
" Refining of Metals	227
" Resistance	5, 112
" " , measurement of	154
" Storage	233–251
" Telegraph, the	310–358
" Transmission, heat developed by	252
" Units	121

INDEX.

	PAGE
Electric Welding	273
Electricity, dynamic defined	1
", relations of, to heat	252–278
", " ", " light	279–309
Electrochemical Action, heat developed by	254
" Equivalence	217
Electrodes and Poles	4
Electrodynamometer, the Weber-Edelmann	152
Electrolysis	206–232
", conditions of	210
" of Mixed Compounds	212
", relations of, to heat	213
", secondary reaction in	211
" of water	10, 209
Electrolyte, relative conditions of current and	219
Electrolytes	206
Electromagnet, the	75
Electromagnets, form of	78
Electromagnetic Poles	75
" Saturation	78
Electromagnetism	71–109
Electrometers	119
Eletcromotive Force	4, 111
" " of discharge in storage cells	246
" ", lowest required in electrolysis	214
" ", Resistance, and Current, units of	6
Electroplating	220
", agitation of the solution	226
", the anodes	221
", auxiliary operations in	223
", plating solutions	223
", required electric energy for	225
", " time of immersion and thickness of deposit	225
", various details	220
Electro-Thermal Capacity of Conductors	255
Electrotyping	226
Element, and Battery, cell	3
Elevated Road Distribution	204
Eleven Year Period, the	50
Energy, electric, conditions of, in the battery cell	4
", ", of improved Faure cell	244

INDEX. 389

	PAGE
Energy, electric, of Planté cell	239
" , " , loss of in the dynamo	195
" " , required for electroplating	225
Equipment, simple line telegraph	314
Equivalence, chemical	216
" , electrochemical	217
Equivalent, Joule's	253
Establishment of the Current, in the arc lamp	299
Exact Observation, of the earth's magnetism	51
Exchange, the telephone	367
Excitation, separate of the dynamo	189
Experiments in Diamagnetism	79
Extra Current	96

F.

	PAGE
Farad, the	118
Faraday's Discoveries, in the electric relations of light	284
" Laws, for electrolysis	215
Faraday, nomenclature by	206
Faradic Current, physiological effects of	107
Faults, location of, in submarine telegraph lines	356
Faure Cell, the	239
" " , " , chemical reaction in	240
" " , " , defects of	240
" " , " improved	241
" " , " " , electric energy of	444
Field, magnetic	60
" Magnets, the, in the dynamo	170, 179
Filament, the, in the incandescent lamp	304
" and Lamp Attachment	305
Forbes' Coulomb-Meter, the	150
Force, deflective, angular measurement of	123
" , electromotive	4, 111
" , magnetic, ascertained by oscillation	42
" , " , " " deflection	42
" , " , lines of	59
" , " , portative	57
" , " , tube of	60
Formation, battery	28
Form of Electromagnets	78
" " Magnets	63

	PAGE
Foucault Currents	56, 196
Fuses, electric	258

G.

Galvani, discoveries of	1
Galvanometer, astatic	129
" , ballistic	134
" , calibration of	124
" , differential	123
" , sine	124
" , tangent	126
" " , the Helmholtz-Gaugain	128
" , Thomson's reflecting	130
Galvanometers	119
" , common	134
Galvanoscope, the	71
Gas Battery, Grove's	233
" Lighting, electric	109
Gauss-Weber Portable Magnetometer, the	69
Generation of Electric Currents by Induction	90
" " Current Dependent on Variation of Intercepted Magnetic Force	95
" , photo-electric	285
" , thermo-electric	259
Generator, the magneto-electric	165
Gonda Leclanché cell	19
Gordon Dynamo, the	185
Gramme Armature, interior wire of the	173
Gravity Ammeters	145
" Cells	25
Grenet Cell, the	16
Grids for storage cell plates, composition of	247
Grotthus' Theory of electrolysis	207
Ground-Switch and Lightning Arrester, cut-out, in the telegraph	319
Grove Cell, the	26
Grove's Gas Battery	233

H.

Hall Process for Aluminium, the	231
Heat Developed by Electric Transmission	252
" " " Electrochemical Action	254

INDEX. 391

	PAGE
Heat, electric, relations of, to electric light	279
" and Light, in the arc lamp	299
" , relations of electricity to	252–278
" , " " electrolysis to	213
Hefner von Alteneck's Regulator	302
Helix of electromagnet	77
Helmholtz-Gaugain Tangent Galvanometer, the	128
History, early, of the telegraph	310
" " , " " telephone	359
Horse-Power, the electric	118
Hughes' Microphone	371
Hunning Transmitter, the, for the telephone	376
Hydrogen Alloy Theory, the, in electric storage	250

I.

Immersion, required time of, in electroplating	225
Improved Commutator	171
" Faure Cell	241
" " " , electric energy of	244
" Transmitters, telephone	362
Incandescent Lamp, the	303
Inclination or Dip	40
Induction, coefficient of magnetic	76
" , " " mutual	95
" Coil	98
" of Core, in coil	97
" " Electric Currents, mutual	84
" , generation of electric currents by	90
" , magneto-crystallic	64
" , results of current	94
" , rotary movement by current	87
" , self	96
Insulation and Clamping, in batteries	9
" " Conductivity	112
Insulator	113
Intensity, absolute magnetic, the earth's	43
Intensity, electric	29, 32
" , magnetic, the earth's	42
Interior Wire of the Gramme Armature	173
International Morse Code, the	312
Interrupter, in the induction coil	99
Inversion, thermo-electric	269

	PAGE
Ions	207
Isoclinic Lines	41
Isodynamic Lines	44
Isogonic Lines	41
Isoclinic Map of the United States	49
Isodynamic Map of the United States	45
Isogonic " " " " "	48

J.

Joule's Equivalent	253
" Law	253
Julien Cell, the, storage	247

K.

Kerr's Discoveries, in the electric relations of light	289
Kündt and Röntgen's Discoveries, in the electric relations of light	289
Key, the telegraph	314

L.

Ladd's Machine	170
Lag, magnetic, in the dynamo	178
Laminated Magnets	56
Lamp, the arc	297
" Attachment, filament and, in the incandescent lamp	305
" , the incandescent	303
Law, Biot's	44
" Cell, the	20
" , Joule's	253
" , Lenz's	93
" , Ohm's	114
Laws, Faraday's, for electrolysis	215
Leclanché Cell, the	17
Lenz's Law	93
Leyden Jar as a Condenser	102
" " and " , the	233
Light, the arc	295
" , Becquerel's discoveries in the electric relations of	288
" , electric, the relations of electric heat to	279
" , Faraday's discoveries in the electric relations of	284
" , the heat and, in the arc lamp	299
" , Kerr's discoveries in the electric relations of	289

INDEX. 393

	PAGE
Light, Kundt and Röntgen's discoveries in the electric relations of	289
", Verdet's discoveries in the electric relations of	287
", magneto-optic polarization	284
", Maxwell's theory of the electric relations of	293
", molecular " " " " " "	294
", photo-electric generation	280
", polarization of	283
", the relations of electricity to	279–309
", strain in the media	294
", summary of the electric relations of	292
Lighting, electric	295
" ", the arc	297
" ", " " lamp	297
" ", " " ", automatic cut-out	303
" ", " " ", " regulation	300
" ", " " ", the carbons	300
" ", " " ", " crater and point	298
" ", " " ", establishment of the current	299
" ", " " ", the heat and light	299
" ", " " ", Hefner von Alteneck's regulator	302
" ", " " ", series distribution	303
" ", " " light	295
" ", " incandescent lamp	303
" ", " " " ", the filament	304
" ", " " " ", filament and lamp attachment	305
" ", " " " . " , multiple series and series multiple distribution	307
" ", " " " ", parallel distribution	305
" ", " " " ", three-wire system	308
Lightning-Arrester, cut-out, ground-switch and, for the telegraph	319
Line, agonic	38
" construction, telegraph	320
" Equipment, simple, in the telegraph	314
Lines of Force, magnetic	59
" " ", isoclinic	41
" " ", isodynamic	44
" " ", isogonic	41
", long distance telephone, transmission on	376
List of Diamagnetic and Paramagnetic Substances	81

	PAGE
Local Action, in the battery	9
Location of the Armature's Magnetic Poles, in the dynamo	177
" " " Poles, in the magnet	63
Locating Faults, in submarine telegraph lines	356
Lodestone, the	35
Long Distance Telephone Lines, transmission on	376
" " Telephony	373
Loss of Energy, in the dynamo	195
" , magnetic, in magnets	56
Lowest Required Electromotive Force, in electrolysis	214

M.

Machine, the Alliance	166
" , Ladd's	170
" , Wilde's	167
Magnet, compensating	74
" , current induced by	90
" , the natural	35
Magnets, artificial	54
" , the field, in the dynamo	170, 179
" , form of	63
" , laminated	56
Magnetic Distribution, lamellar	63
" Effects, in electrolysis	216
" and electric phenomena, analogy between	67
" Field	60
" Force Ascertained by Deflection	42
" " " " Oscillation	42
" " , generation of current dependent on variation of intercepted	95
" Induction, coefficient of	76
" Intensity, the earth's	42
" " , absolute, the earth's	43
" Lag, in the dynamo	178
" Lines of Force	59
" Loss, in magnets	56
" Maps	40
" " , of the hemispheres	39
" Penetration	63
" Polarity	35
" Poles, the earth's	37

		PAGE
Magnetic Poles, location of the armature's, in the dynamo		177
" Saturation		55
" Shells		63
" Storms		50
" Strength, in the electromagnet		76
Magnetism		35–70
" , Ampere's theory of		89
" , coefficient of		98
" , origin of terrestrial		44
" as a Mode of Molecular Motion		65
" , residual		57
" , terrestrial, illustrated		41
Magneto-Crystallic Induction		64
Magneto-Electric Generator		165
Magneto-Optic Polarization		284
Magnetometer, the Gauss-Weber, portable		69
Make and Break, in the electric circuit		93
Map, agonic, of the United States		52
" , isoclinic," " " "		49
" , isodynamic, of the United States		45
" , isogonic, " " " "		48
Maps, magnetic		40
" , " , of the hemispheres		39
Mariner's Compass, the		35
Maxwell's Theory of the electric relations of light		293
Measurement of Angles		122
" " Deflective Force, angular		123
" , electric		110–164
" " , different kinds of		118
" of " Resistance		154
Media, strain in the, in the electric relations of light		294
Megohm, the		116
Mercuric Bisulphate Cell, the		17
Metals, electric refining of		227
Microfarad, the		118
Microvolt, "		116
Microphone, Hughes'		371
Milliammeter, the Weston		138
Milliampere, the		117
Milliken Repeater, the, in the telegraph		326
Mixed Compounds, electrolysis of		212
Molecular Motion, magnetism as a mode of		65

	PAGE
Molecular Theory of the electric relations of light	294
Morse Code, the American	312
" ", " International	312
Motor, the alternating current	198
", development of the electric	190
", the dynamo as a	193
", " " and	165–205
", principles of the	193
" as a converter, the Tesla	202
", the Westinghouse Tesla	199
Motors, series, shunt, and compound wound	196
", thermo-magnetic	204
Multiple Arc defined	29
" Series and Series Multiple distribution	308
" Switch-Board, the telephone	367
Multiplier, the Schweigger	72
Multiplex Telephony	372
Mutual Induction of Electric Currents	84
" ", coefficient of	95

N.

	PAGE
Natural Magnet, the	35
Needle, the astatic	73
", " dipping	40
" Telegraph, the	311
Neutral Relay, the telegraph	342
Nobili's Rings	216
Nomenclature by Faraday	206

O.

	PAGE
Observation, exact, of the earth's magnetism	51
Ohm, the	6, 116
Ohm's Law	114
One-Fluid Cells	13–22
" " ", polarization of	22
Open Circuit defined	19
Operation of Condenser	102
" " the Polar Duplex	338
" " " Quadruplex, construction and	341
" " " Voltaic Cell	6
Operations, auxiliary, in electroplating	223

	PAGE
Ores, electric reduction of	228
Origin of Terrestrial Magnetism	44
Oscillation, magnetic force ascertained by	42

P.

Pacinotti-Gramme Armature, the	170
Parallel Distribution, in electric lighting	305
Paramagnetic and Diamagnetic Bodies	64
Peltier Effect, the, in the electric relations of light	268
Penetration, magnetic	63
Perforation, electric	107
Period, the eleven year	50
Periods, secular	46
Photo-Electric Generation	280
" " Reduction of Resistance in Selenium	282
Photophone, the	282
Physiological Effects of Faradic Current	107
Pile, dry, Zamboni's	7
", the Voltaic	2
Planté's Secondary Cell	236
Plates for Faure cell, electric preparation of the	244
" " Planté ", " " " "	236
" " storage cells, effects of charge and discharge on	245
Plating Solutions	223
Point, the crater and, in arc light carbons	298
Polar Attraction and Repulsion	58
" Duplex, the	333
" ", ", operation of	338
Polarity, magnetic	35
Polarization	9
" of One-Fluid Cells	22
" of Light	283
" , magneto-optic	284
Polarized Relay, the	336
Polarizer, the, in the polarization of light	284
Pole-Changer, the	334
Poles, consequent, in the magnet	61
", the earth's magnetic	37
", electrodes and	4
", electromagnetic	75
", location of the, in magnets	63

INDEX.

	PAGE
Position of the Brushes, in the dynamo	178
" and current, of the incandescent lamp	305
Portable Magnetometer, the Gauss-Weber	69
Portative Force, magnetic	57
Potassium Bichromate Cell	15
Potential, electric	110
Power, distribution of	203
Pressure, electric	111
Primary Circuit, current induced by opening or closing	93
" Current, " " " varying the strength of	94
Principles of the Motor	193
" " " Telephone	360
Printing Telegraphs	358
Prism or Gonda Leclanché Cell	19
Pumpelly Cell, the	247

Q.

Quadruplex, construction and operation of the	341
" , repeating by the	347
" Telegraphy	340
Quantity electric	29, 31, 32

R.

Rapid Transmission, the Wheatstone system of automatic, in telegraphy	348
Reaction, chemical, in the Faure cell	340
" , " , " " Planté "	337
" , secondary, in electrolysis	211
Recorder, siphon, Thomson's	353
Reduction of Ores, electric	228
Refining of Metals, "	227
Reflection, effects of double, on magneto-polarized light	292
Reflecting Galvanometer, Thomson's	130
Register, the telegraph	315
Regulation, automatic, in the arc lamp	300
Regulator, Hefner von Alteneck's	302
Relations of Electricity to Heat, the	252–278
Relations of Electricity to Light, the	279–309
" " Electric Heat to Electric Light, the	279
" " Electrolysis to Heat	213

	PAGE
Relay, the telegraph	317
" " " differential	333, 336
" " " neutral	342
" , polarized	336
Rheostat, water	101
Relative Conditions of Current and Electrolyte, in electrolysis	219
" Time of Charging and Discharging storage cells	249
Repeater Connections, telegraph	327
Repeater, the button	325
" , " Milliken	326
Repeaters, telegraph	324
Repeating by the Quadruplex	347
Repulsion, polar attraction and	58
Required Electric Energy, in electroplating	225
" Time of Immersion and Thickness of Deposit in electroplating	225
Residual Magnetism	57
Resistance Coils	155
" , electric	5, 112
" , " , measurement of	154
" in Selenium, photo-electric reduction of	282
Results of Current Induction	94
Reversal, effect of current, in electrolysis	217
" of Rotation, in the motor	202
Reversible " " " "	197
Rings, Nobili's	216
Rotary Movement by Current Induction	87
Rotation, reversal of, in the motor	202
" , reversible , " " "	197
Rules, Ampere's	73
Ruhmkorff's Commutator	104

S.

San Francisco, secular variation at	53
Saturation, electromagnetic	78
" , magnetic	55
Schewigger Multiplier, the	72
Secondary Cell, Planté's	236
" Reaction, in electrolysis	211
Secular Periods	46
" Variation	46

	PAGE
Secular Variation at San Francisco	53
" " in the United States	48
" " at Washington	51
Selenium, photo-electric reduction of resistance in	282
Self-Induction	96
Separate Excitation of the dynamo	189
Series Distribution in electric lighting	303
" multiple, multiple series and, distribution	307
" , Shunt, and Compound Winding, in the dynamo	180
" , " , " " Wound Motors	196
Shells, magnetic	63
Shunt and Compound Winding, Series	180
Siemens Armature, the	167
Sign, battery	3
Signaling Apparatus, for the telephone	366
Silver Chloride Cell, the	27
Simple Line Equipment, telegraph	314
Sine defined	122
" Galvanometer	124
Siphon Recorder, Thomson's	353
Sliding Core, in the induction coil	101
Smee's Cell	13
Solenoid, the	83
Solution, agitation of, in plating	226
Solutions, plating	223
Sounder, the telegraph	316
Spark, "	97
" Coil	109
Special Construction, in induction coils	102
Station Arrangement, telegraph	321
Stearns Duplex, the	330
Storage Cell, composition of grids for plates	247
" " , conductivity and buckling in	246
" " , effects of charge and discharge on the plates	245
" " , preparation of the plates	244
" " , E. M. F. of discharge	246
" " , the Faure	239
" " , " " , chemical reaction in	240
" " , " " , defects of	240
" " , hydrogen alloy theory	250
" " , Improved Faure	241

INDEX. 401

	PAGE
Storage Cell, Improved Faure, electric energy of	244
", ", ", ", " preparation of the plates	244
", ", the Julien	247
", ", Planté's, electric energy of	239
", ", ", chemical reaction in	237
", ", ", electric preparation of the plates	236
", ", the Pumpelly	247
" Cells, capacity of	249
", ", durability of	249
", ", relative time of charging and discharging	249
", ", weight of	247
Storage, electric	233–251
Storms, magnetic	50
Strain in the Media, in the electric relations of light	294
Strength, magnetic	76
Submarine Telegraphs	352
Substitution of the Dynamo for the Battery, in the telegraph	347
Summary of the relations of electricity to light	292
Surveyor's Compass, the	37
Suspension, bifilar	129
Switch-Board, the telegraph	324
", ", " telephone multiple	367
System of automatic rapid transmission, the Wheatstone	348
" " long distance telephony, the American	375
" " " " ", Van Rysselberghe's	374
", three-wire, in electric lighting	308

T.

Table, Ampere's	82
" of Thermo-Electric Potential of Metals	264
Tangent defined	125
" Galvanometer	126
" " , the Helmholtz-Gaugain	128
Telegraph, the electric	310–358
" " " , the American Morse code	312
" " " , " battery	314
" " " , " button repeater	325
" " " , cut-out, ground-switch, and lightning arrester	319
" " " , construction and operation of the quadruplex	341

INDEX.

	PAGE
Telegraph, the electric, the dial	357
" " ", early history of	310
" " ", the international Morse code	312
" " ", " key	314
" " ", line construction	320
" " ", locating faults in submarine lines	356
" " ", the Milliken repeater	326
" " ", needle	311
" " ", operation of the polar duplex	338
" " ", the polar duplex	333
" " ", " polarized relay	336
" " ", " pole-changer	334
" " ", printing telegraphs	358
" " ", the register	315
" " ", " relay	317
" " ", repeater connections	327
" " ", repeaters	323
" " ", repeating by the quadruplex	347
" " ", simple line equipment	314
" " ", the sounder	316
" " ", station arrangement	321
" " ", the Stearns duplex	330
" " ", substitution of the dynamo for the battery	347
" " ", switch-board	323
" " ", the Wheatstone system of automatic rapid transmission	348
Telegraphs, printing	358
" , submarine	352
Telegraphy, diplex transmission in	340
" , duplex	329
" , quadruplex	340
Telephone, the	359–377
" ", accessory apparatus	365
" " the Bell	361
" ", early history of	359
" ", the exchange	367
" ", Hughes' Microphone	371
" ", multiple switch-board	367
" ", principles of	360
" ", signaling apparatus	366

INDEX. 403

	PAGE
Telephone, the, theory of telephonic transmission	371
" ", transmission on long distance lines	376
" transmitter, the Blake	364
" " , " Edison	363
" " , " Hunning	376
" , the, improved transmitters	362
Telephonic Transmission, theory of	371
Telephony, long distance	373
" , " " , the American system	375
" , " " , Van Rysselberghe's system	374
Multiplex	372
Terrestrial Magnetism Illustrated	41
" " , origin of	44
Tesla Motor as a Converter, the	202
" " , the Westinghouse	199
Theory of Electric Generation in the Cell	6
" " Grotthus, in electrolysis	207
" , the hydrogen alloy, in electric storage	250
" of Magnetism, Ampere's	89
" , Maxwell's, of the electric relations of light	293
" , the molecular of the electric relations of light	294
" of Telephonic Transmission	371
Thermo-Electric Diagrams	266
" " Generation	259
" " Inversion	269
" " Potential of Metals, table of	264
Thermo-Magnetic Motors	204
Thermopile, the	270
Thickness of Deposit, in electroplating	225
Thomson Effect, the, in the electric relations of heat	269
Thomson's Reflecting Galvanometer	130
Three-Wire System of distribution, in electric lighting	308
Time of Immersion and Thickness of Deposit, in electroplating	225
Torsion Balance, Coulomb's	67
Transformer, or converter, in the alternating current system	190
Transmission, diplex, in telegraphy	340
" , heat developed by electric	252
" on Long Distance Telephone Lines	376
" , theory of telephonic	371
" , the Wheatstone system of automatic rapid	348
Transmitter, the Blake telephone	364

	PAGE
Transmitter, the Edison telephone	363
" , " Hunning "	376
Transmitters, improved "	362
Tube of Force	60
Two-Fluid Cells. Battery Formation	23–34
" " " , construction of	23

U.

Units, electric	121
" of Electromotive Force, Resistance and Current	6

V.

Vacuo, discharge in air and in	107
Van Rysselberghe's System of long distance telephony	374
Value of Volta's Discoveries	3
Variations, annual and diurnal, in terrestrial magnetism	50
" , cosmic " " "	51
" , secular " " "	46
" , " , at San Francisco	53
" , " , in the United States.	48
" , " , at Washington	51
Various Details of electroplating	220
Verdet's Discoveries, in the electric relations of light	287
Vibrator, in induction coil	99
Volt, the	6, 116
Volt-ampere	118
Volta, discoveries of	2
Volta's " , value of	3
Voltaic Battery, the. Definitions	1–12
" Cell, operation of the	6
" Pile, the	2
Voltameter, the water	151
Voltameters	151
Voltmeter, the Cardew	147
" , " Weston	135
" , " Wirt	139
Voltmeters and Ammeters	134
" " " , Ayrton & Perry's spring	142

W.

Walker's Cell	14
Washington, secular variation at	51

	PAGE
Water, electrolysis of	10, 209
" Rheostat	101
" Voltameter, the	151
Watt, the	118
Weber-Edelmann Electrodynamometer, the	152
Wheatstone Bridge, the	157
" System of Automatic Rapid Transmission, the	348
Weight of Storage Cells	247
Welding, electric	273
Westinghouse Dynamo, the	186
" Tesla Motor, the	199
Weston Ammeter, the	138
" Milliammeter, the	138
" Voltmeter, "	135
Wilde's Machine	167
Winding electromagnets	76
" , series, shunt, and compound	180
Wirt Voltmeter, the	139

Z.

Zamboni's Dry Pile	7, 8
Zinc, amalgamation of the, in battery cells	8
Zinc-Carbon Cells	13

Southern Electrical Supply Co.,   Gate City Electric Co.,
823 LOCUST ST.,   522 DELAWARE ST.,
ST. LOUIS, MO.   KANSAS CITY, MO.

Central Electric Company,

116 & 118 FRANKLIN STREET, CHICAGO.

MANUFACTURERS, DEALERS, AND IMPORTERS OF ALL KINDS OF

Electrical Supplies.

TRADE MARK.

GENERAL WESTERN AGENTS OKONITE WIRE AND PRODUCTS.

IMPROVED CANDEE WIRE.
   THE BUTLER HARD RUBBER GOODS.
     THE CELEBRATED PACKARD LAMP.
       THE BRADY MAST ARM, ETC., ETC.

We carry at all times a complete assortment of the latest improved electrical specialties, and solicit correspondence.

CONNECTED BY PRIVATE WIRE WITH POSTAL TELEGRAPH-CABLE COMPANY.

Electric Power Transmission Co.,   Western Electrical Supply Co.,
1722 LAWRENCE ST.,   418 S. FIFTEENTH ST.,
DENVER, COLO.   OMAHA, NEB.

FACTORIES: ANSONIA, CONN.

MANUFACTURERS AND DEALERS IN

GENERAL ELECTRICAL SUPPLIES

WESTERN AGENTS FOR

A HIGH GRADE OF RUBBER INSULATION.

SOLE MANUFACTURERS OF

MOISTURE-PROOF LINE WIRE.

SUNBEAM LAMPS AT SUNBEAM PRICES.
EDISON LAMPS AT EDISON PRICES.

LIST OF WORKS
ON
ELECTRICAL SCIENCE
PUBLISHED AND FOR SALE BY

D. VAN NOSTRAND COMPANY,

23 Murray and 27 Warren Streets, New York.

ATKINSON, PHILIP. The Elements of Electric Lighting. Including Electric Generation, Measurement, Storage and Distribution. Sixth edition. 104 illustrations, 260 pages. 12mo, cloth.................................... $1 50

—— Elements of Static Electricity, with full description of the Holts and Topler Machines and their mode of operating. 65 illustrations. 12mo, cloth..................... 1 50

BADT, F. B. The Dynamo Tender's Hand-Book. With 70 illustrations. 16mo, cloth................................ 1 00

—— Incandescent Wiring Hand-Book. With 41 illustrations and five tables. Second edition. 12mo, cloth........... 1 00

—— Bell-Hanger's Hand-Book. 97 illustrations. 12mo, cloth 1 00

BOTTONE, S. R. Electrical Instrument-Making for Amateurs. A Practical Hand-book. With 48 illustrations. Fourth edition. Enlarged by a chapter on The Telephone. 12mo, cloth. Reduced to 50

—— The Dynamo, How Made and How Used. A Book for Amateurs. Sixth edition, with additional matter and illustrations. 39 illustrations 1 00

—— Electric Bells and all about them. A Practical Book for Practical Men. With more than 100 illustrations. 12mo, cloth. Reduced to... 50

—— Electro-Motors. How Made and How Used. A Handbook for Amateurs and Practical Men. Many illustrations. 12mo, cloth..... .... 1 20

CUMMING, L. Electricity Treated Experimentally. For the use of Schools and Students. New edition. 12mo, cloth.. 1 50

DU MONCEL, COUNT. Electro-Magnets. The Determination of the Elements of their Construction. 16mo, fancy boards. (Van Nostrand's Science Series, No. 64.). $0 50

FISKE, Lt. BRADLEY, A., U. S. N. Electricity in Theory and Practice; or, The Elements of Electrical Engineering. Eighteenth edition. 8vo, cloth. 180 illustrations....... .... 2 50

HASKINS, C. H. The Galvanometer and its Uses. A Manual for Electricians and Students. Fourth edition, revised. 12mo, morocco, illustrated................... 1 50

HEAP, Major D. P. Electrical Appliances of the Present Day. Being a report on the Paris Electrical Exhibition of 1881. 8vo, cloth, fully illustrated..... ... 2 00

HOBBES, W. R. P. The Arithmetic of Electrical Measurements. With numerous examples fully worked. 12mo, cloth................................. 50

HOPKINSON, DR. JOHN. Dynamic Electricity: Its Modern Use and Measurement, chiefly in its Application to Electric Lighting and Telegraphy; including—I. Some Points on Electric Lighting. II. On the Measurement of Electricity for Commercial Purposes. By T. N. Shoolbred. III. Electric Light Arithmetic. By R. E. Day. 16mo, boards. (Van Nostrand's Science Series, No. 71.)....... . 50

INCANDESCENT ELECTRIC LIGHTING. A practical description of the Edison system by L. H. Latimer, to which is added a description of the Edison Electrolytic Meter and a paper on the maximum efficiency of Incandescent Lamps. By John W. Howell. (Van Nostrand's Science Series, No. 57.) 16mo, boards.......... 50

INDUCTION COILS. How Made and How Used. (Van Nostrand's Science Series, No. 53.) 16mo, boards.. 50

KAPP, GISBERT. Electric Transmission of Energy and its Transformation, Subdivision, and Distribution. A practical Hand-book. Third edition. 8vo, cloth... ..... 3 00

—— Alternate Current Machinery. Illustrated. 16mo, boards. 50

KEMPE. The Electrical Engineer's Pocket Book: Modern Rules, Formulæ, Tables and Data. 32mo, leather. . 1 75

LORING, A. K. A Hand-book of the Electro-Magnetic Telegraph. Second edition. 18mo, boards............ 50

LOCKWOOD, T. D. Electricity, Magnetism, and Electric Telegraphy. A Practical Guide and Hand-book of General Information for Electrical Students, Operators, and Inspectors. 8vo, cloth, 376 pages, 152 illustrations...... 2 50

MONROE AND JAMIESON. Pocket Book of Electrical Rules and Tables. For the use of Electricians and Engineers. Seventh edition, revised. 32mo, leather. Adapted for the pocket.................................. $2 50

NIPHER, Prof. F. E. Theory of Magnetic Measurements. With an appendix on the Method of Least Squares. 12mo, cloth....................................... 1 00

NOAD, H. M. The Student's Text-Book of Electricity. A new edition, carefully revised by W. H. Preece. 8vo, cloth, illustrated..................................... 4 00

POPE, F. S. The Modern Practice of the Electric Telegraph. New edition. 8vo, cloth......................... 1 50

PLANTE, GASTON. The Storage of Electrical Energy, and Researches in the Effects Created by Currents Combining Quantity with High Tension. With 89 illustrations. Translated from the French by Paul B. Elwell. 8vo, cloth... 4 00

SALOMONS, SIR DAVID. Electric Light Installations and the Management of Accumulators. A Practical Hand-book. Fifth edition, revised and enlarged. With 99 illustrations. 348 pages, 12mo, cloth................ 1 50

SCHELLEN, Dr. H. Magneto-Electric and Dynamo-Electric Machines: Their Construction and Practical Application to Electric Lighting and the Transmission of Power. Translated from the third German edition, with large additions and notes relating to American Machines, by N. S. Keith. Vol. I., with 353 illustrations. 8vo, cloth...... 5 00

THOMPSON, SILVANUS P. Dynamo-Electric Machinery. A Series of Lectures, with an introduction by Frank L. Pope. 16mo. Numerous illustrations, boards. (Van Nostrand's Science Series, No. 66.)................ 50

—— Recent Progress in Dynamo-Electric Machines: being a Supplement to Dynamo-Electric Machinery. 16mo. (Van Nostrand's Science Series, No. 66.)................ 50

WATT, A. Electro Deposition: A Practical Treatise on the Electrolysis of Gold, Silver, Copper, Nickel and other Metals and Alloys. 12mo, cloth, illustrated............. 3 50

—— Electro-Metallurgy Practically Treated. New and enlarged edition. 12mo, cloth........................... 1 00

WALKER, FREDERICK. Practical Dynamo Building for Amateurs. 18mo, boards........................ 50

** *A General Catalogue of Works in all branches of Electrical Science furnished on application.*